Applied Biology/ Chemistry

Natural Resources

Written by
Alan C. Kousen
and
John J. Roper

Applied Biology/Chemistry is a project sponsored by a consortium of State Vocational Education Agencies and developed by the Center for Occupational Research and Development with the assistance of science educators.

Published and distributed by:
CORD Communications
324 Kelly Drive
Waco, Texas 76710
817-776-1822   FAX 817-776-3906

Printed in USA   April 93

ISBN 1-55502-363-0 (Applied Biology/Chemistry)
ISBN 1-55502-364-9 (Natural Resources)

Preface

The world is full of things we humans didn't put here: air, water, soil, plants, animals, and fuels, such as coal and oil. We depend on these natural resources, as they are called, for survival.

Nevertheless, most of us are removed from the world of nature. Water comes to us by way of the faucet; our air is "conditioned" to cool it; the only plants and animals we know well are houseplants and housepets; and few of us ever lay eyes on the fuel that keeps electricity humming through our household appliances.

When we do hear about natural resources, we hear about problems. For example, without even trying, most of you probably have become aware of acid rain, the loss of valuable rain forests, and the need to "save the whales."

How do you deal with so much bad news? It can be discouraging, and you may feel like ignoring it. But it's important to realize that this is your planet, too. The way we use resources—the way we preserve them and protect them—is part of everybody's life. This unit is about how our survival depends on air, water, soil, fuels, plants, and animals.

You'll find out about how different resources affect one another and affect us. For example, you'll discover how burning coal and oil can pollute the air, creating acid rain, which in turn pollutes rivers and streams and affects plant and animal life. You also will begin to understand the *positive* way that resources interact. Plant life provides needed oxygen to the air; animals die and return valuable minerals to the soil; some animals eat plants, others eat other animals, humans eat both animals and plants. By the time you get to the end of this unit, you may begin to wonder if anyone or anything can ever operate with total independence!

This unit also features occupations in which people take an active part in protecting and using natural resources. You'll meet a power plant operator, a water treatment lab technician, a businessman who develops and sells soils, and many others.

Table of Contents

NATURAL RESOURCES

UNIT GOALS

After you complete this unit, you will be able to —

1. Decide whether or not a natural resource will be available in the future.

2. Give examples of how natural resources are used to produce energy, make products, provide food and shelter, and improve the quality of life.

3. Analyze problems that result from obtaining and using natural resources.

4. Propose solutions to problems resulting from obtaining and using natural resources.

5. Relate jobs to natural resources.

INTRODUCTION TO NATURAL RESOURCES

THINK ABOUT IT

- What is the most popular type of tennis shoe worn on your campus? Why is it so popular?

- What materials are used to make this tennis shoe?

- Do you think that we are in danger of running out of any of these materials?

SUBUNIT OBJECTIVES

After you complete this subunit, you will be able to—

1. Identify and give examples of natural resources.
2. Classify natural resources by the following categories:
    a. Limited resource
    b. Unlimited resource
    c. Renewable resource
    d. Nonrenewable resource
3. Explain what depletion and degradation mean in terms of natural resources.
4. Analyze jobs to see how natural resources are involved in every job.

LEARNING PATH

To complete this subunit, you will—

1. View and discuss the video, "Introduction to Natural Resources."
2. Read the text through "Could We Use Up Our Natural Resources?"
3. Take part in class discussions and activities.
4. Do Laboratory 1, "How Is a Resource Depleted?"
5. Read the remainder of the subunit text.
6. Take part in class discussions and activities.

What Are Natural Resources?

Imagine

- You are the first person to set foot on a planet like Earth in a galaxy far away. It is Earth's twin planet in all ways, except it has never had people on it.

- You're the only person there.

Be Aware

- What do you see, hear, and feel immediately around you? Above and below you?

- What do you find beyond your immediate surroundings?

ACTIVITY 1-1

- In a special Applied Biology/Chemistry (ABC) notebook, make a list of at least 10 things you would expect to find on this twin planet. Keep in mind that it's just like Earth, except that it has no people.

- Compare your list with lists made by other students. Add things you may have overlooked. You will use this list in a later activity.

The things on your list are natural resources. They are the things that existed on Earth before people became a part of the picture. These include air, water, soil, plants, and animals. Natural resources include both living and nonliving things.

If you should stay on this far-away planet, these natural resources would mean life or death to you. You would use them to provide food, clothing, and shelter. If a group of people should colonize this planet, they would use these resources for transportation and communication.

ACTIVITY 1-2

- Draw a six-foot, imaginary circle around where you're sitting now.

- In your ABC notebook, list everything within that circle that has been made by people.

- For each thing, identify the natural resources that were required to make it. Don't forget to include what you're wearing on the list. If you don't know the natural resource used to make something, ask another student or your teacher.
- Create a master list for the class.
- What are the most commonly used resources?

Throughout this unit, you will be presented with situations and job descriptions based on real-life events and people. Some of these situations will relate to a particular workplace; others will focus on community or personal problems. You will be asked to read about the situation (in the indented, italic print) and to address problems or answer questions about it.

In this unit, one of the situations you will read about is the Edison Power company. Although the Edison Power Company does not exist, the story is based on interviews with employees of an actual power company.

Edison Power

Edison Power Company is an electric utility. Its business is producing electricity. Edison has plans to build a new power plant to serve Richmond, a city of about 300,000. Several years ago, the city began to attract new industry. A number of new businesses and manufacturing companies have moved there. Since Richmond has undergone rapid growth, its energy needs have increased significantly.

The new Edison Power plant will consist of two generating units. Each unit will have a maximum capacity of 600 megawatts. A megawatt is one million watts. This plant will meet the energy needs of Richmond and the surrounding area for the foreseeable future.

Edison Power Company intends to create Lake Edison by damming Big Creek, a spring-fed tributary of the Crystal River. Then they will build the power plant next to the lake. The company has obtained a lease on a bed of lignite coal (a low-grade coal) in the area. This means that its source of fuel will be mined nearby. Figure 1-1 is a map of the area where

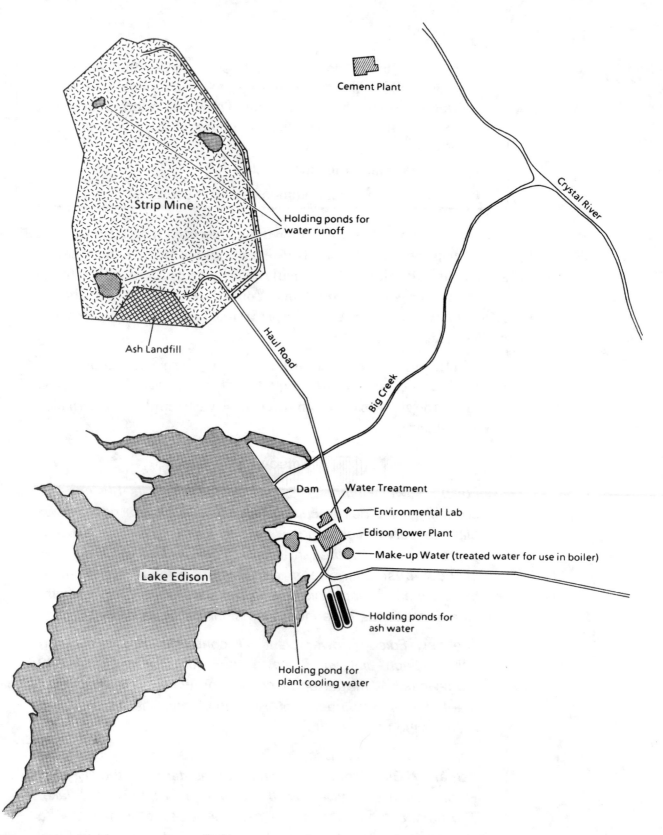

Figure 1-1
Map of the proposed Edison Power Company power plant and strip mine

the power plant and mine will be. Since the coal will have to be transported only a few miles to the plant, it will cost less. Also, it's cheaper to use coal than to use petroleum.

The process of producing electricity at a power plant is shown in Figure 1-2. The plant operation depends on the use of several natural resources. First, the power plant boilers burn coal, which generates heat. This heat is used to change water into steam. The steam drives a turbine, which turns a generator to produce electricity. Water is also used to cool the steam and condense it back to water. The condensed water is then reused to produce steam.

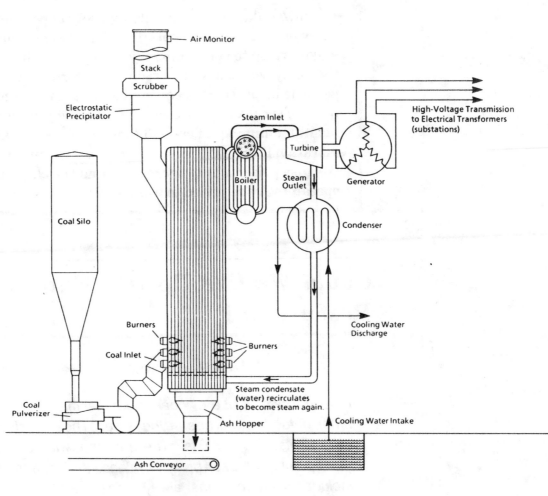

**Figure 1-2
Power
plant**

The plant and mining operation combined will provide jobs for about 700 people in the local community. Edison Power Company will contribute over $3,500,000 a year in taxes to the local community.

JOB PROFILE: POWER PLANT OPERATOR

James W. has been a power plant operator for a power plant in Detroit for almost twenty years. He says that his job can be "very quiet, very routine for days or even weeks, but then—watch out! Everything happens at once in a power plant because the plant operations are interconnected. You have to stay alert for that one little reading that's off, that one event that tells you you have a problem. Then you have to react fast. You have to know the system." What does he like the most about the job? He gives a big smile and says, "All that power."

Power plant operators fill a variety of positions. At the entry level, they may operate fuel- and ash-handling systems and move up to operating major support equipment. Finally, they may work as control room operators. They must make periodic inspections of equipment, put equipment in and take it out of service, operate systems, clean equipment, and monitor a number of critical temperatures, oil levels, and vibration levels. Operators are generally union workers. Because power plants operate around the clock, power plant operators are required to work shifts.

Could We Use Up Our Natural Resources?

Edison Power & Natural Resources

Edison Power's use of coal and water has a big effect on the other natural resources in the area. These include air, water, soil, plants, and animals. The area that will be affected by the new plant includes both farmland and ranchland. It also includes land that has always been left in its natural state. Natural vegetation includes pines, oaks, and native grasses. Wildlife includes many species of birds, coyotes, fish and game animals.

People in the region of the plant have asked the power company to do all it can to protect this environment. Edison Power Company wants to cooperate. It plans to develop a model operation for mining, producing electricity, and reclaiming the land after it has been mined. Edison Power is establishing an environmental advisory committee. This committee will help plan the power plant and mining operation. It also will monitor and evaluate the process for reclaiming the land throughout the life of the plant.

Our lives literally depend upon the Earth's supply of natural resources (Figure 1-3). It stands to reason, then, that we are concerned about their availability. Is the supply of natural resources great enough to last? If we should totally use up a vital resource, how would we get along without it? Could we replenish the supply?

**Figure 1-3
Natural
resources**

Limited and Unlimited Resources

A resource that can be used up is a limited resource. At any given time, the quantity of a limited resource can be measured or estimated. Every time we use a limited resource, the quantity that exists is decreased. If our use of a limited resource continues indefinitely, at some point that resource will disappear from Earth. Coal is an example of a limited resource, one that may one day be used up. Let's look at Edison Power Company's use of coal.

How Long Will Edison Power Company's Coal Last?

Edison Power Company knows that the lignite deposit they've leased next to the power plant site is limited. It will be used up someday. The company needs to know how long it can depend upon this lignite as a fuel.

Scientists called geophysicists have estimated the amount of lignite at the site. They estimate that the bed of lignite contains 240 million tons of coal. The plant will burn about 6 million tons of coal each year.

ACTIVITY 1-3

- Answer the following questions in your ABC notebook.

  1. How long will the mine be able to provide the fuel required by the power plant? Make your calculations in your notebook.

  2. Do you think that the coal will last long enough for the plant's electricity production to pay for the investment?

  3. What information might you need to answer this question?

Air is another natural resource that's important to Edison Power Company. Why? Because it's needed to burn coal. Nothing burns without air. Edison, however, isn't concerned about the quantity of air that exists. Why is this?

The supply of air is very different from the supply of most other natural resources. Unlike coal, for example, it is not likely to be used up. No matter how much air we use, the supply doesn't seem to decrease. There's always more. Air, then, is an example of an unlimited resource. Can you think of other unlimited resources?

Nonrenewable and Renewable Resources

Even though a resource may be limited, it doesn't necessarily follow that we will run out of that resource. Consider trees, for example, any kind you might think of—oak, maple, pine, spruce. There are a lot of trees on Earth. But at any particular time, only so many trees exist. Therefore, they are a limited resource.

Trees and coal are both limited resources. But trees do something that coal doesn't. They grow and reproduce in a relatively short time. By contrast, coal takes about a million years to form.

When a resource can produce more of itself to replace what has been used, it is called renewable. Trees, then, are a renewable resource (Figure 1-4). A resource that cannot produce more of itself, such as coal, is a nonrenewable resource.

**Figure 1-4
Trees, a
limited,
renewable
resource**

ACTIVITY 1-4

Refer to the lists of resources you made for Activities 1-1 and 1-2.

- Classify each natural resource on your lists in two different ways—as renewable or nonrenewable and as limited or unlimited.

- Make a table that lists limited/unlimited, renewable/ nonrenewable resources.

- Compare your table with those of other students and discuss the differences.

What Problems Do We Face with Natural Resources?

Using natural resources often creates problems. In general, there are two types of problems regarding natural resources: (1) problems regarding the **quantity** of natural resources and (2) problems regarding the **quality** of natural resources.

The Problem of Quantity: Depletion

The problem of quantity is the problem of depletion. A natural resource is being depleted when using it reduces the amount that exists. A limited, nonrenewable resource such as petroleum someday may be totally depleted. An unlimited resource, such as air, can last forever (although its quality can be degraded).

As a resource is depleted, it may become more and more difficult to obtain. This is true of aluminum. In West Virginia, for example, most of the rich veins of aluminum have been mined. The veins being mined now have much less aluminum than the veins formerly mined. This makes getting the aluminum more difficult—and more expensive.

Even renewable resources can be depleted. This can happen if we use the resource faster than it can renew itself. A particular type of plant, for example, can be depleted, even though it's a renewable resource. For example, mahogany, a popular wood for furniture, comes from trees that grow in tropical forests. Today these forests are being cut in vast stretches. Trees like mahogany are not able to renew themselves when all the trees in a large area are cleared. Cutting the forest disturbs the soil so much that mahogany seeds won't grow to produce more trees. So the world's supply of mahogany is being rapidly depleted.

The Problem of Quality: Degradation

When we lower the quality of a natural resource, we degrade it. Unfortunately, using a natural resource often lowers the quality of that resource. For example, when you wash your clothes, you use water. When you dispose of the dirty water, you contribute to water pollution.

Sometimes, the use of one resource may lower the quality of some other resource. When you drive your car, you burn gasoline made from petroleum. Burning the gasoline contributes to air pollution. Burning coal in a power plant also contributes to air pollution.

JOB PROFILE: ENVIRONMENTAL AND CHEMICAL ANALYSIS TECHNICIANS

Gary R. and Tina M. are Environmental and Chemical Analysis Technicians (ECATs) who work for Edison Power in a plant similar to the proposed Richmond plant. Gary's primary responsibility is for air quality; Tina's is for water quality.

The main by-product of the power plant operations is ash: bottom ash (heavy enough to fall to the bottom of the stack) and fly ash (light enough to float out into the atmosphere). Fly ash emissions must be kept within certain limits established by the Environmental Protection Agency (EPA) and the state air quality control agency.

Gary's job is to check the air quality equipment and monitoring instruments. If fly ash emissions go over or even near the EPA limit, Gary tracks down the problem and corrects it. Gary is a fisherman and outdoorsman who takes great pride in his work. "This area is my home." he says, "It's important to me to protect the quality of the environment."

Tina oversees water quality for the plant. She supervises the disposal of bottom ash into large ponds that are lined with clay to prevent the ash from contaminating the ground water or the land around it. She also monitors the water that is released from the cooling water system back into the lake. This effluent, as it is called, is kept in holding ponds until its temperature drops close to that of the lake. Both Tina and Gary spend a lot of time preparing routine reports for the EPA and the state.

Like Gary, Tina discovers and corrects problems when they occur. "But the name of the game for us," says Tina, "is prevention. We don't wait for problems. We anticipate them and stop them from happening."

We must use natural resources. Life without them would be impossible. Our challenge is to learn how to use them while degrading them as little as possible.

Problems related to natural resources are discussed throughout this unit. These problems are among the greatest problems that our society faces as we move into the twenty-first century.

ACTIVITY 1-5

- Contact an environmental and chemical analysis technician at your local power plant for a telephone interview or a visit to the class.

- Make a list of natural resources the ECAT monitors and protects.

- Make a list of duties the ECAT performs on a daily and weekly basis.

- Find out what kind of training or education the ECAT had to have to get the job.

How Do Our Jobs Relate to Natural Resources?

ACTIVITY 1-6

- Pick a job that interests you. It may be a job one of your parents, relatives, or friends holds. It may also be a job that you plan for yourself in the future.

- In your ABC notebook, write a brief description of your chosen job. Include the tasks performed by people who hold the job and the skills needed to perform each task.

- Then write about how this job relates to natural resources. Refer to the list given just after this activity. Your teacher may have you report to the class about your job.

You might not think that most jobs are connected to natural resources. Nevertheless, every occupation is involved with natural resources in one way or another. Seven of these ways are shown in the list below.

1. **Maintaining Resources** – Example occupations: aquatic biologists and wildlife managers. They protect and conserve natural resources and help both plants and animals reproduce and grow.

2. **Collecting or Harvesting Natural Resources** – Example occupations: miners, fishers, farmers. They gather natural resources so that they can be used as materials for products, as food, or as sources of energy.

3. **Making Products** – Example occupations: auto manufacturing workers, furniture makers, food processing workers. They make products from natural resources.

4. **Converting Natural Resources to Energy** – Example occupation: power plant operators. They may burn coal or petroleum in power plant boilers to produce electricity.

5. **Using Products** – Example occupations: child care workers, cosmetologists, journalists, interior designers. All jobs use products made from natural resources.

6. **Providing Services** – Example occupations: truckers who transport products made from natural resources, salespeople who sell them. These people are involved in the delivery of natural resources to users.

7. **Protecting and Using Natural Resources Wisely** – Example occupations: water treatment technicians, air pollution monitors, conservationists, teachers, lawyers, nurses, doctors, nutritionists. Protecting and using natural resources wisely cut across all occupational areas and involve people in many different types of jobs.

ACTIVITY 1-7

- Interview someone whose job is directly related to natural resources.

- Write the results of the interview in your ABC notebook. Your teacher may assign a group to put all interviews together into a career resource book for the class. Your teacher may suggest questions to ask that will help you have a successful interview.

What Is the Value of Natural Resources?

The value of natural resources often is difficult to determine. For a farmer or rancher, the cost of land changes over time. Lumber companies estimate the value of timber before harvesting, but the final market price may go up or down. The price of oil changes according to how much oil is sold worldwide.

It is even harder to figure out the worth of resources that are not bought or sold. What is the value of clean air in a city, or a lake with clear, safe water? What is the value of plants or animals that we don't use?

ACTIVITY 1-8

It is fairly easy to know the value of natural resources when they are found in products—leather in a pair of shoes, for example, or the metal in a new car.

- Discuss the worth of resources that are not used to make products. Think about the resources described below.
  - 200 acres of forest where wildlife, which is becoming rare in the area, lives.
  - 10 miles of undeveloped oceanfront (one-half mile wide) with sandy beaches—forty-five minutes away from the center of a city of 300,000 people.
  - A clean, clear, spring-fed swimming hole (1000 ft by 2000 ft) on five acres of land near the center of town.

- Survey the opinions of others in your class in regard to developing areas such as those described below.
  How would you feel about opening all these areas to extensive development if—
    — Plants and animals in the area would lose their homes?
    — It is over 200 miles to the next undeveloped natural area?
    — Many more people would be able to enjoy them?
    — Many people would get good jobs after the areas were developed?
    — The development activity would result in millions of additional tax dollars for your city?
    — You could get rich from selling all or some portion of these areas for extensive development?
  What restrictions would you place on the development?

Looking Back

Natural resources are all those things that exist on Earth without help from people. Such resources include air, water, soil, plants, and animals. They can include living and nonliving things. We depend on the Earth's supply of natural resources and are very concerned about their availability. Some resources are **limited** and can be used up, like coal. Others are **unlimited** and are not likely to be used up, like air. Some resources are **renewable**, such as trees which take a relatively short time to grow. Others are **nonrenewable** like coal, which takes millions of years to form. **Depleting** natural resources reduces the amount that exists. **Degrading** natural resources lowers their quality, which happens in the case of water pollution.

Any job that you consider depends somehow on natural resources. Our challenge is to use these resources wisely.

Vocabulary

The words and phrases below are important to understanding and applying the principles and concepts in this subunit. If you don't know some of them, find them in the text and review what they mean. They're listed in the order in which they appear in the subunit.

natural resource	nonrenewable resource
limited resource	depletion
unlimited resource	degradation
renewable resource	harvesting

Further Discussion

- What are the main natural resources available in your area? Which is the most abundant? Which is the least available?

- Are there jobs in your area related to the conservation and management of natural resources? If possible, ask a representative to come to your class to tell about his/her job.

Activities by Occupational Area

General

The Effect of Depletion and Degradation of Natural Resources on Jobs

Discuss how the depletion and degradation of natural resources has changed jobs in the local community in the last 40 years. What local resources have been depleted during this time? What local resources have been degraded during this time? Have these environmental changes caused any occupations to disappear? Have these environmental changes caused changes in any occupations? Have these environmental changes created any new occupations?

School Cleanup Day

Organize a school cleanup day. If possible solicit contributions from local businesses for this effort. Notify TV and radio stations and newspapers to get news coverage.

News Broadcasts

Do "news broadcasts" on local environmental issues from your perspective. Use a video camera to film the broadcasts. Write your own scripts from research based on actual news stories from TV, newspapers, journals, etc.

Prepare a "futuristic news broadcast" based on how you think environmental issues will impact the future.

Agriculture and Agribusiness

Deforestation

The tropical rain forests in many countries are being clear-cut to make the land available for the production of other crops. How does this practice affect the global environment? What effect does this practice have on the local soil? What can be done to correct this situation?

Photo of Local Farm

Get a photograph of a local farm. From the photo, identify different elements in the scene. Classify each element as natural or man-made, limited or unlimited, and renewable or nonrenewable.

Health Occupations

Organ Transplantation

The development of organ transplant surgery has created a new resource. Discuss donated organs as a resource. What ethical problems can arise from the abuse of this resource? How can these problems be avoided?

Health Problems Caused by Environmental Degradation

Discuss different health problems that are caused by or made worse by environmental degradation. What can be done to correct the environmental degradation associated with these problems? What can be done to treat the problems?

Home Economics

Identification of Natural Resources in Consumer Products

Collect several consumer products such as shampoo, hairspray, potato chips, bath soap, room deodorizer, breakfast cereal, etc. Use product labels to identify the natural resources used in each product. Summarize the findings in a list that includes the consumer product, the ingredients, and the natural resources used for the ingredients.

Trash Activity

Get a bag of trash from your teacher. The bag will have items such as: a soft drink can, shoe, chip package, tissue, etc. Also get a card that says renewable, nonrenewable, limited or unlimited.

Go through the bag of trash considering the following questions.

a. Which items are limited and renewable?

b. Which items are limited and nonrenewable?

c. Which items are unlimited and renewable?

d. Which items are unlimited and nonrenewable?

Individually go through the bag of trash and categorize each item. Then, as a group discuss the category of each item and reach a consensus. Are there any patterns? Are there examples for a, b, c, and d? Why?

Did your group get the same results as the other groups in the class?

Answer the following questions.

1. What natural resources are found in the bag of trash?

2. Describe which resources in the trash are available forever. Which resources are in danger of running out?

3. Could any of these trash items be recycled? How does recycling help the supply of natural resources?

4. What problems are created when trash is discarded? Which of your items will easily break down? Which won't?

5. List some jobs associated with the creation, use, and disposal of this bag of trash.

6. What can you and your family do to help preserve natural resources through proper trash disposal?

Industrial Technology

Natural Resources Used in the Auto Industry

List the limited resources used in the automotive industry. Which of these materials are recyclable? Suggest substitutions for limited resources that can't be recycled.

Recycling

Discuss local industries that benefit from recycling. What natural resources are recycled? Are there natural resources that are not recycled that could be? Are there any industries that are not recycling resources that could benefit from it?

HOW IS A RESOURCE DEPLETED?

Introduction

One of the most important aspects of using natural resources is that the total amount of any resource is finite. Therefore, a nonrenewable resource can be steadily harvested only a limited amount of time before it is depleted.

Resources, however, are rarely used up steadily over time, as you might use up a tube of toothpaste. The demand for products made from natural resources may vary greatly through time. This demand is closely linked to how much users must pay for products. And, the price of products depends on how much raw material is left and how costly it is to obtain. For instance, when the supply of petroleum was believed to be dwindling in the 1970s, the price of petroleum products, especially gasoline, rose sharply.

Natural resource managers follow patterns of resource production and consumption. Using this information, they predict if and when a natural resource will be depleted. In this lab you will apply one of the chief tools of resource managers, a type of graph known as the resource depletion curve.

Purpose

In this lab, you will observe the typical pattern of natural-resource depletion.

Lab Objective

When you've finished this lab, you will be able to —

- Plot a resource depletion curve for a simulated mining operation.

Lab Skill

You will use this skill to complete this lab —

- Generate and graph a set of data.

Materials and Equipment Needed

large shoe box

birdseed or dried corn, 5 lb

soybeans or other small white beans, 8 oz

4 paper cups

stopwatch, timer, or wall clock with second hand

graph paper

4 plastic teaspoons

400-ml beaker

large container

4 forceps (optional)

4 blindfolds (optional)

**LAB
PROCEDURE**

Pre-Lab Discussion

During the mining of metals, ores are extracted from the earth. However, because not all of the ore is used, it must be refined to obtain the desired metal. For instance, copper ore is mined and then refined to obtain the copper. The leftover, or gangue, is usually disposed of or returned to the earth.

Mining of copper ore begins in a rich vein of copper. As more and more copper ore is extracted from the vein, the concentration of copper in the remaining ore decreases. As the copper concentration decreases, the mining effort must increase to produce the same quantity of usable copper.

In this lab, you will simulate a mining operation for a useful mineral. As you proceed, you will observe the typical pattern of resource depletion.

Method

Each group of students is a mining company. Each mining company is assigned to one mine (shoe box).

1. Fill the shoe box 50% full with seed.

2. Add the beans to the shoe box and mix thoroughly.

In the remaining steps (the mining operation), several techniques are possible. Either the teacher will tell you which method to use or the group will vote on the best method for its mine. You will use one of the following techniques:

- Forceps—one bean at a time (simulates using a pick axe)

- Thumb and forefinger—one bean at a time (simulates using pneumatic tools)

- Plastic teaspoon—miner wearing blindfold (simulates dynamiting or bulldozing)

- Plastic teaspoon—without blindfold (simulates an engineer or geologist using high-tech equipment)

After ore is removed from the mine, it should be placed in the ore car (paper cup).

3. Mine three rounds (three years) as follows: **Note:** Each miner represents one shift (or crew) of miners.

    Year 1 (one miner)--Mine for 15 sec.

    Year 2 (two miners)--Mine together for 15 sec.

    Year 3 (three miners)--Mine together for 15 sec.

    Rotate the miners so that each group member mines one or two times during the first three years. Continue the rotation through the end of the last round so that the work load is distributed equally among the group members.

4. Following each round of mining, refine the ore as follows:

    - Pour ore onto sheet of paper.

    - Extract the beans from the seed (or feed) and count the total number of beans picked. (Each bean represents one ton of metal.)

    - Store the refined mineral (beans) in your paper cup until the end of each round. Then pour it into the stock pile (the 400-ml beaker).

    - Dispose of the gangue by pouring it into the disposal container.

    - Assign one group member to graph the results as you complete each round. (Figure L1-1 shows you how to set up this graph.)

5. Graph the mining results in your ABC notebook. (Refer again to Figure L1-1.)

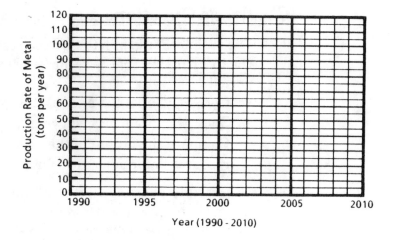

**Figure L1-1
Graph paper
for resource
depletion
curve**

6. Report the results of your group's mining effort to the teacher, so that they can be added to the Class Data Table on the board.

7. Beginning with year 4, repeat Steps 4, 5, and 6 as many rounds as time permits, three miners mining together each round. If in any year a miner gets no beans, he/she must quit mining.

WRAP-UP

Conclusions

Refer to your completed resource depletion curve to answer the following questions:

1. In which year did production of metal increase at the fastest rate? What factors at work in the mining industry accounted for this rapid increase?

2. In which year did production fall most sharply? What factors accounted for this sharp fall?

3. Based on your curve, how many tons of metal remained in the mine after you completed the final year of mining?

4. If the fall-off in production continues at the current rate, in what year will the metal in your mine be depleted? Using a dotted line, extend your curve downward from the last year of actual mining to show this total depletion.

Challenge Questions and Extensions

5. Create another graph (using the results of this lab) that shows the total amount of the resource consumed over the mining period.
   - Correctly label the axes of this graph.
   - Show the totals for each year.
   - Compare the height of the new curve to the area under the resource depletion curve for different years.

6. Use a triple-beam or digital balance to estimate the amount of metal (in tons) currently remaining in the mine. (You must plan the correct procedure and carry out the appropriate measurements.)
   - Compare this estimate to the one you obtained from the resource depletion curve.
   - How can you explain the difference?

7. Attempt to plan a new mining procedure that would increase your company's profits by creating a slower depletion curve? Explain why such a plan is possible (or impossible).

8. Plan a workable recycling program or other nonmining approach for obtaining useful minerals.

9. How has modern technology helped slow down the depletion of copper and other minerals? (Think in terms of new materials and applications that are making some uses of copper and other minerals obsolete.)

FOSSIL FUELS

THINK ABOUT IT

- Can you think of any connections between sports cars and the times when dinosaurs lived?

- Do you think sports cars are in danger of being extinct like dinosaurs? Why?

SUBUNIT OBJECTIVES

After you complete this subunit, you will be able to —

1.   Identify uses of fossil fuels.

2.   Describe the chemical composition of fossil fuels.

3.   Explain the combustion of fossil fuels as a chemical reaction.

4.   Analyze major problems of using fossil fuels.

5.   Evaluate possibilities for replacing fossil fuels with alternative sources of energy.

6.   Identify jobs related to fossil fuels.

LEARNING PATH

To complete this subunit, you will—

1.   View and discuss the first half of the video, "Fossil Fuels: A True Story."

2.   Read text through "How Do We Get Energy from Fossil Fuels?"

3.   Take part in class discussions and activities.

4.   Do Laboratory 2, "How Do We Measure the Energy Stored in a Fossil Fuel?"

5.   View and discuss the second half of the video, "Fossil Fuels: A True Story."

6.   Read the remainder of the subunit text.

7.   Take part in class discussions and activities.

FOSSIL FUELS

Fossil fuels are found beneath the Earth and the ocean floor. They are called fossil fuels because they formed from plant and animal materials that lived millions of years ago. We get each type of fossil fuel—petroleum, natural gas, and coal—from the Earth by using different methods of recovery.

Since petroleum is a liquid in its natural state, it can be extracted from wells, as water is. Natural gas often is found in the same place as petroleum. Natural gas escapes from a well through a pipe. Then, it's directed into pipelines.

Mining Coal for Electric Power Plant Fuel

Lignite is the fuel used at Edison Power. The mine site is 14,000 acres. The coal beds are located between 30 and 140 feet from the surface of the ground. When Edison Power evaluated mining methods, they looked at deep-shaft mining, auger mining and surface mining, all described below.

Deep-shaft mining *– Large shafts are drilled down from the surface of the ground to the coal seam, and then tunnels follow the coal seam.*

Auger mining *– A large pit is dug down beside the seam of coal. Then large drills or augers are used to bore into the coal seam. Enough space must be left between the holes in the seam to prevent cave-ins. This means that you can't recover all of the coal from the seam.*

Surface mining (also called strip mining) *– All of the rock and soil above the coal seam is removed. Then the coal is removed and the pit is filled in with the original material. This method is usually the most economical for coal seams within 200 feet of the surface.*

ACTIVITY 2-1

You are a reclamation engineer with the Edison Power Company. You will be responsible for restoring the land to a usable state after it has been mined.

- Based on the information given, decide which mining method you would recommend for the site described. Why?

- How do you think this operation will affect natural resources in the area?

Write the answers in your ABC notebook.

Why Are Fossil Fuels Important to Us?

Fossil fuels are important natural resources. They include petroleum, natural gas, and coal. We use fossil fuels primarily in two important ways: as a source of energy and as raw materials for making other products. Figure 2-1 graphically shows what percents of coal and petroleum are used for energy and for products.

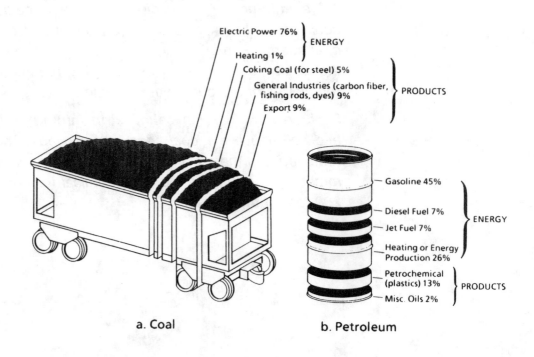

Electric Power 76% } ENERGY

Heating 1%

Coking Coal (for steel) 5%

General Industries (carbon fiber, fishing rods, dyes) 9% } PRODUCTS

Export 9%

Gasoline 45%

Diesel Fuel 7%

Jet Fuel 7% } ENERGY

Heating or Energy Production 26%

Petrochemical (plastics) 13% } PRODUCTS

Misc. Oils 2%

a. Coal

b. Petroleum

**Figure 2-1
Uses of
fossil fuels**

ACTIVITY 2-2

- Answer the following questions in your ABC notebook.
  1. Based on the graphs in Figure 2-1, what percent of coal is used for generating energy? What percent of petroleum is used for energy?
  2. What is the largest category of products that is derived from petroleum?
  3. What building material is derived from coal?
  4. Make a list of the ways you depend on fossil fuels for energy. For products.

Fossil Fuels as a Source of Energy

Energy usually is defined as the ability to do work. In today's world, most of the energy needed to do work comes from the chemical energy of fossil fuels. You use fossil fuel as a source of energy each time you ride in a gasoline-powered automobile. Fossil fuels also provide the energy we use to keep houses and buildings warm in the winter. Fossil fuels power factories, airplanes, and ocean liners. They are also the fuel most commonly used to generate electricity.

Fossil Fuels as a Source of Raw Materials

Although the major use of fossil fuels is for energy, that's not the only reason fossil fuels are important to us. When you walk on a vinyl floor or drink from a Styrofoam cup, you're using fossil fuels, or petroleum. Vinyl and Styrofoam are synthetic materials made from fossil fuels. Synthetic materials are chemical compounds produced in factories and processing plants.

To make synthetic materials from fossil fuels, chemicals are first derived from petroleum. These chemicals are called petrochemicals. Petrochemicals are the basis of many industries. Plastics, synthetic fibers, fertilizers, paints, food additives, packaging materials, automobile parts, and thousands of other products come from petrochemicals.

What Is the Chemical Composition of Fossil Fuels?

Each of the fossil fuels—petroleum, natural gas, and coal—is really a mixture of hydrocarbons. Hydrocarbons are compounds made from atoms of the elements carbon and hydrogen.

Figure 2-2 shows the structuring of methane, the simplest hydrocarbon.

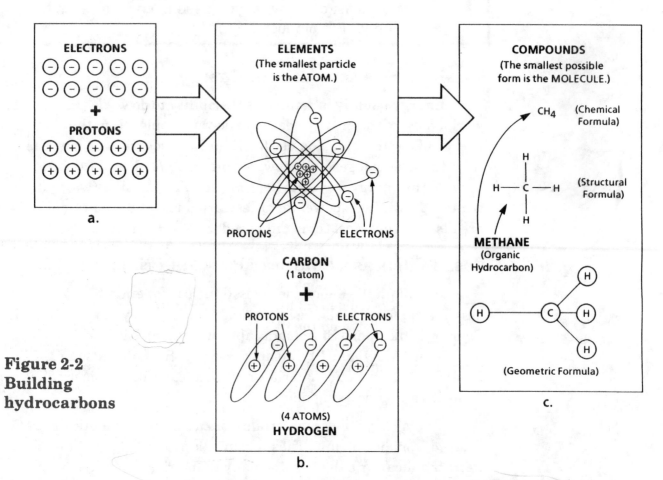

**Figure 2-2
Building
hydrocarbons**

Every atom consists of a central nucleus surrounded by electrons. The nucleus contains positively charged particles called protons. The protons attract an equal number of negatively charged particles called electrons. (See Part A of Figure 2-2.) For example, the nucleus of a carbon atom has six protons; therefore, it attracts six electrons. The electrons are found at various distances outside the nucleus. An atom is the smallest form of an element that has all the properties of that element. (See Part B of Figure 2-2.)

Elements are the basic substances that make up the world. They also determine the characteristics of all materials. Atoms of elements join together to form a molecule. The molecule is the smallest possible form of any compound, but not all molecules are compounds. Compounds are substances that have two or more different elements. (See Part C of Figure 2-2.)

Hydrocarbons are organic compounds. In organic compounds, carbon atoms form the center, or backbone, of the molecule. Each type of hydrocarbon has its own unique molecular structure.

All fossil fuels have hydrocarbons in them. Coal, however, is primarily pure carbon, with only a few hydrocarbons interspersed throughout. High grades of coal have a higher percent of pure carbon than low grades. (Elements such as sulfur also may be bound in coal. Sulfur may be found in both high- and low-grade coals.)

The major hydrocarbon found in natural gas, a fossil fuel, is the molecule methane. The methane molecule is represented in Part C of Figure 2-2 in three ways: structural, chemical and geometric. In the structural and geometric formulas, you can see that a carbon atom is the center of the molecule.

In a methane molecule, each hydrogen atom is held to the carbon atom because its single electron is attracted by the nucleus of the carbon. One of the carbon's four outermost electrons is likewise attracted by the nucleus of hydrogen. These attractions represent a chemical bond between the two atoms. This type of bond involving the sharing of electrons is a covalent bond. The bond stores chemical energy as a result of the attractions between electrons and nuclei. The more energy that's stored in the chemical bonds of a molecule, the more energy there is to be released during a chemical reaction.

Besides methane, natural gas also includes propane, butane, and several similar molecules. The structural and chemical formulas for propane are shown below. You can see that carbon atoms form the backbone of the molecule.

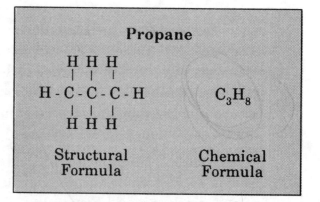

Propane

$$H - C - C - C - H$$

(with H atoms above and below each carbon)

C_3H_8

Structural Formula   Chemical Formula

The mixture we call natural gas may have different proportions of methane (CH_4) and propane (C_3H_8), depending on its source. It also may have other hydrocarbons such as ethane and butane.

ACTIVITY 2-3

Butane is a fossil fuel that you might burn in an outdoor grill. Butane is very similar to propane, except that it has one more carbon atom in its backbone and two more hydrogen atoms.

- Write the structural formula for butane. (Use the formula for propane as a guide.)
- Write the chemical formula for butane. How many chemical bonds are in a single butane molecule?

How Much Energy Is Stored in Different Fuels?

For simple hydrocarbons, the number of bonds in a molecule is directly related to the amount of energy stored in that molecule. Hydrocarbon fuels that are gases are normally sold by volume. (A volume of any gas contains the same number of molecules as the same volume of any other gas.) Burning a liter of methane gas releases approximately 9500 calories. You can figure out the calories released by a liter of propane in the following way:

$$\frac{10 \text{ bonds in propane}}{4 \text{ bonds in methane}} = 2.5$$

Since a molecule of propane has two and a half (2.5) times the number of bonds as a molecule of methane, propane can release two and a half times the number of calories when burned. Therefore, multiply the caloric output of methane by 2.5.

2.5 × 9500 calories/liter = 23,750 calories/liter of propane

ACTIVITY 2-4

Use the calculation above as a model to figure out the number of calories released when a liter of **butane** is burned.

JOB PROFILE: GAS ANALYSIS TECHNICIAN

Gas analysis technicians usually work for energy companies or power plants. They can work both in the field and in the laboratory. Various duties might include taking samples of gases for component analysis, adjusting pressure and temperature indicators, and calculating relative densities. The technician might also test for hydrogen sulfide gas, and calculate volumetric rates and heating values.

Both field and laboratory technicians require high school diplomas. The lab technician should have some college chemistry. A two-year degree in chemistry is ideal for both positions.

How Do We Get Energy from Fossil Fuels?

We can use fossil fuels for energy because of one very important characteristic: they burn. When they burn, they release energy in the form of heat. This burning is called combustion.

What Is Combustion?

- Combustion is a chemical reaction in which a fuel combines rapidly with the oxygen in the air.

- A chemical reaction is a process in which a substance is changed into one or more new substances.
- Some chemical reactions break down compounds into elements or smaller compounds.
- Other chemical reactions join elements or compounds together to make new compounds.

What Happens During Hydrocarbon Combustion?

Hydrocarbon combustion is the technical term that refers to burning fossil fuels. Burning hydrocarbon fuels breaks chemical bonds in the molecules of the fuel source. Then oxygen in the air forms new bonds with the hydrogen and carbon atoms from the fuel.

Figure 2-3 summarizes the combustion reaction of fossil fuels. The reactants are the elements and compounds that combine in a reaction. Products are the compounds and elements that are formed during a reaction. Energy also has an important role in chemical reactions. The energy content of the reactant molecules is different from the energy content of the product molecules. This energy difference causes the reaction to absorb energy from the surroundings or to release energy in the form of heat or light.

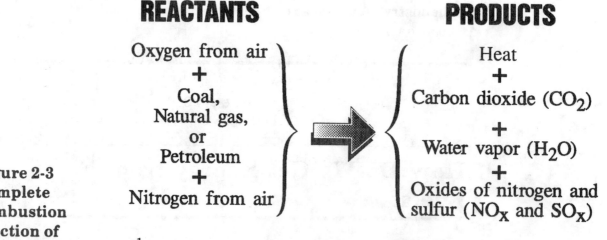

REACTANTS

Oxygen from air
+
Coal,
Natural gas,
or
Petroleum
+
Nitrogen from air

PRODUCTS

Heat
+
Carbon dioxide (CO_2)
+
Water vapor (H_2O)
+
Oxides of nitrogen and sulfur (NO_X and SO_X)

[1]"X" can represent any number from one to three

Figure 2-3
Complete
combustion
reaction of
fossil fuels

Applied Biology/Chemistry

ACTIVITY 2-5

- Research and answer these questions in your ABC notebook or in class discussion.

  1. What is the most useful product of hydrocarbon combustion?
  2. What is the least desirable product of hydrocarbon combustion?
  3. Where do the nitrogen and sulfur that form oxides come from?
  4. What gas would be formed if the amount of oxygen combined with fuel were not enough to form carbon dioxide?

You can obtain information about the burning of fossil fuels by calling your local power plant or by looking up fossil fuels in a science encyclopedia.

ACTIVITY 2-6

You are a sophomore at Richmond High. You will use the electricity produced by the planned lignite plant. Figure out how much coal must be burned each year to supply you and your family with electricity.

Here are two facts that you will need to solve this problem—

- One kilowatt is approximately equal to 860,000 calories of energy (a calorie is a unit of energy).

- The lignite has approximately 4,000 calories of energy per gram (or 4,000,000,000 calories per ton). The energy per gram of lignite is the heat content for lignite. Each fuel has its own heat content.

What other information or records do you need in order to solve this problem? Show your calculations in your ABC notebook.

It's interesting to note that most of our foods are combinations of carbon and hydrogen, along with a few other elements. This makes them behave as a sort of "fuel" for our bodies. Like fossil fuels, food combines with oxygen. Either reaction releases energy. Also, each reaction produces carbon dioxide (CO_2) and water. Of course, the reaction in the body does not involve fire, and it produces only small amounts of heat.

What Problems and Issues Are Related to Fossil Fuels?

Fossil fuels are both limited and nonrenewable. By current estimates, Earth's total supply of fossil fuel reserves will last for several centuries. They will become increasingly expensive, though, as we use up the most available sources. Also, some fossil fuels may become totally unavailable in some areas. Edison Power, for example, knows that the supply of lignite near Richmond will last only about 40 years. Since the reserves of oil, coal, and natural gas are not evenly distributed around the world, international politics also affect the availability of the fuels. As supplies decrease, it's important to look for ways of using them more efficiently while we look for other sources of energy.

ACTIVITY 2-7

- Identify the major industries in your area and answer these questions about them.
  1. How do they use fossil fuels?
  2. How would the industries and the occupations your parents work in be affected by fossil fuel shortages?
  3. How might fuel shortages affect your community?
- Report your findings to your class.

Besides availability, there are other problems with fossil fuels. Petrochemicals and the chemicals used to process them sometimes find their way into our drinking water. Many plastics and other synthetic materials made from petrochemicals pollute our waterways and fill up our landfills. Since their chemical composition is very stable, these materials may exist for hundreds of years before natural processes break them down into compounds that can be absorbed back into the land and water.

When we burn fossil fuels, we have more problems. As you've already learned, you get more than heat when you burn fossil fuels. You also get by-products: water vapor, CO_2, and oxides of nitrogen and sulfur. Some of these by-products create problems. They contribute to the problem known as smog. They also contribute to two less obvious problems: acid rain and global warming. These problems will be discussed in more detail in the next subunit, "Air and Our Atmosphere."

What Can Be Done About the Problems?

Some alternatives to fossil fuels do exist. We're learning how to make synthetic materials and fuels from plants, for example. With our current technology, however, we couldn't replace fossil fuels entirely.

You know that fossil fuels are a major source of energy. You might ask why we don't use more electricity for our energy instead of fossil fuels. The fact is, we generate most of our electricity by using fossil fuels. Heat energy from burning fossil fuels is converted into electricity. So using more electricity isn't the answer—unless we can generate it without using fossil fuels. Other possible sources of energy include nuclear fission, solar (the sun), geothermal (heat inside the earth), and wind.

ACTIVITY 2-8

- Divide the class into teams of three or four and let each group investigate these alternative sources of energy:
    — Nuclear fission
    — Nuclear fusion
    — Solar power
    — Geothermal energy
    — Wind power
    — Tidal generation
  What are the advantages and disadvantages of each type?
- Find out how the electricity for your area is produced and answer these questions.
    1. If a fossil fuel is used, what is it?
    2. What is the source of the fuel?
    3. How does the fuel get to the power plant?
    4. How long is fuel from this source expected to last?
- Share your findings with your class.

Nuclear Fission

Nuclear energy has been proposed as a major alternative to fossil fuels. The center of an atom, its nucleus, contains energy. In a nuclear power plant, the energy contained in uranium (U) is released by splitting the nuclei of uranium atoms. This process is called nuclear fission, and it creates large amounts of heat. This heat, like that produced by burning fossil fuels, can be used to produce steam. The steam then turns a turbine, which powers an electrical generator. A single nuclear power plant can generate enough electricity to meet all the energy needs of a large city.

The use of nuclear power poses other problems. Nuclear fission creates new elements as waste products. These new elements are highly unstable and radioactive. That is, they emit streams of tiny high-energy particles (subatomic particles) as they change to a more stable form. This high-energy radiation is harmful to living things. Radioactive wastes must not be allowed to enter the environment. Instead, they must be buried in containers deep in the earth and safeguarded for thousands of years. An affordable way of doing this on a large scale has not yet been found.

Another problem with nuclear power is that uranium is a limited, nonrenewable resource.

Solar Power

There are several ways to use the sun as a source of energy. These ways include passive solar heating, active solar heating, and photovoltaic energy.

Passive Solar Heating. The use of the sun to provide energy is not new. For years, homes have been built with most of the windows on the southern side to provide maximum sunlight in the living areas. (In the northern hemisphere, where we live, the sun is always angled toward the south.) This use of sunlight is called passive solar heating (Figure 2-4). It has little impact on the environment, and it helps to decrease reliance on fossil fuels for heating. Passive solar, however, never completely replaces the need for a secondary heat source, such as an electric heater or oil furnace.

Active Solar Heating. Sunlight also can be used to produce heat in a more active way. Active solar heating requires the use of special solar collectors positioned on the roof of the house or very close to it (Figure 2-4). The sun heats a liquid being circulated through pipes in an insulated black housing. This heated liquid, usually an antifreeze solution, then is pumped to a storage tank. Here, its temperature is checked. If necessary, additional heat may be supplied by oil, gas, or electricity. This hot liquid can be used to heat water. The hot water produced can be used as the hot water supply for the house, or it can heat the house. When used to heat a house, the hot water is pumped through radiators, or it heats air that is blown through the house.

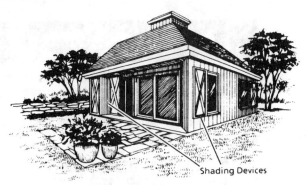

Passive Solar Heating
(large window faces south)

Active Solar Heating

Figure 2-4 Passive and active solar energy

JOB PROFILE: OWNER OF SOLAR ENERGY COMPANY

As energy costs have risen, harnessing solar energy has become more popular. Many solar energy businesses have been created. An owner of a small solar energy company might sell solar energy systems to residential and commercial clients. The needs of the client would have to be analyzed, and then a system would be designed and installed. Additionally, the small businessperson has to manage all aspects of the business, including hiring personnel, bookkeeping, maintaining inventory, advertising, etc. Communication skills are very important in business, as the businessperson will be working with a wide variety of people.

ACTIVITY 2-9

- Collect at least six cans of similar size. Soft drink cans will work well.
- Paint two of the cans white and two black. Leave two unpainted.
- Fill the cans with water.
- Measure the temperature of the water in each can. The temperature should be the same in all of the cans at the start.
- Place one white, one black and one unpainted can in a window where they will be in direct sunlight. Place the remaining cans where they are out of direct sunlight.
- In your ABC notebook, record the temperature of the water in each can every 5 minutes for one class period.
- Make a chart to organize your data.
- When you complete your recording, make a graph of the temperatures.
- Discuss your results as a class.

Photovoltaic Energy. Another form of solar energy is called photovoltaic energy. Certain materials, like cadmium sulfide (CdS), produce electricity when they are exposed to light. These materials are made into flat plates with electrical leads attached to them. Sometimes, they are called "solar cells."

Unfortunately, the current technology for photovoltaic energy is not very efficient. It takes many of these small cells to produce enough electricity to do even a little work. Calculators work very well on photovoltaic energy. Toasters do not. On a large scale, photovoltaic energy is not economical.

Geothermal Energy

Water beneath the Earth's surface sometimes is superheated by Earth's internal heat. This produces steam. This steam can be piped to buildings for heating or for driving machinery. The steam also can turn turbines to produce electricity. This type of energy is called geothermal energy. Geothermal energy is relatively nonpolluting, clean, and safe.

There are only so many places where geothermal energy can be tapped. Among these are New Zealand, Iceland, and some parts of California. Each of these places uses geothermal energy to one degree or another.

Wind Power

Wind power is becoming an attractive alternative for many people. Large towers are erected to carry small, wind-powered generators (Figure 2-5). Electricity produced by the wind-powered generators may be sold to power companies. This lets power companies cut back on the amount of electricity that is generated using fossil fuels. Of course, on days without much wind, the wind turbines are less effective.

If wind turbines were used extensively, they would create a host of new problems. They make noise, block views, and disturb animal habitats. Clearly, no energy solution is perfect.

Figure 2-5
Wind
turbines

Newer Energy Technologies

Tidal Generation. New energy technologies are being studied all the time. Getting energy from the ocean is a possibility in areas where there's a large difference between the levels of the high and low tides. This is called tidal generation. At this time, a few demonstration plants are producing small amounts of energy from the rise and fall of the oceans.

Nuclear Fusion. In nuclear fusion, the nuclei of two small light atoms fuse into one heavier one. In the fusion process, energy is released. This is the way the sun produces its heat and light. The problem is to get the nuclei to join, or "fuse," because they both are positively charged and tend to repel each other. For nuclear fusion to be useful as a source of energy, the reaction must produce more energy than is required to cause the reaction. Scientists have been conducting research for over thirty years to find a way to get more energy out of the reaction than goes in.

Early fusion reactors took 10 million times more energy to force the nuclei together than was released. Over the past 20 years the energy required has been reduced greatly, but it still takes more energy than is released. Research continues on other techniques to produce fusion. It may be that within your lifetime fusion will become a feasible power source. The benefits would be tremendous. Fusion plants would generate less radioactive waste than today's nuclear fission power plants. Fusion produces almost none of the air pollution that a power plant produces when fossil fuels are burned.

ACTIVITY 2-10

Since using fossil fuels presents some serious problems, why do we use them? We know that some fossil fuels will become very expensive. Someday, they will run out. So, what are we going to do then? Do we have any alternatives?

- Find the list you made in Activity 2-2 of ways we use and depend on fossil fuels.
- Add to the list if you now see other uses.
- Make a check by those uses that you think are the most important.
- Place an X next to those uses you think we could do without.
- For each "most important" use that you checked, describe a way to meet the need without relying on fossil fuels.

Looking Back

Fossil fuels include petroleum, natural gas, and coal. They are important natural resources. They are a mixture of **hydrocarbons**, compounds made from atoms of the elements **carbon** and **hydrogen**. When hydrocarbons burn, they react with oxygen and release energy as chemical bonds are broken. This energy is in the form of heat.

Most of the world's energy supply comes from the chemical energy of fossil fuels. Some of the by-products of burning fossil fuels create problems, such as smog, acid rain, and global warming. Alternatives to burning fossil fuels include nuclear energy, solar power, geothermal energy, and wind power.

Vocabulary

The words and phrases below are important to understanding and applying the principles and concepts in this subunit. If you don't know some of them, find them in the text and review what they mean. They're listed in the order in which they appear in the subunit.

fossil fuel	combustion
synthetic material	chemical reaction
petrochemical	chemical reactant
chemical composition	chemical product
hydrocarbon	oxide
compound	heat content
atom	nuclear fission
element	solar energy
nucleus	passive solar energy
electron	active solar energy
proton	photovoltaic energy
molecule	geothermal energy
chemical bond	tidal generation
calorie	nuclear fusion

Further Discussion

- Your teacher will divide the class into several groups. Each group chooses a problem concerning fossil fuels. Make an overhead transparency that graphically illustrates your group's problem. Be ready to suggest solutions to your problem.

- As a class, make a list of all jobs you can think of in your community that are directly related to fossil fuels.

- Discuss the effect that insulating a home has on fuel bills.

Activities by Occupational Area

General

Construct Molecular Models

Use gumdrops and toothpicks and construct molecular models of methane (CH_4), propane (C_3H_8), and butane (C_4H_{10}). The butane molecule can have all 4 carbons in a straight chain (n-butane) or it can be branched (iso-butane).

Field Trip to Fossil Site

Take a field trip to a local site with various fossils. Invite a local "rockhound" to go along and relate the formation of the fossils to the formation of fossil fuels.

Agriculture and Agribusiness

Agricultural Products as Substitutes for Fossil Fuels

Discuss the pros and cons of using agricultural products to produce ethanol as a substitute for fossil fuels. Discuss the effect on air quality of each type of fuel. Discuss other products that are produced now from fossil fuels (different polymers) that might be made from agricultural products.

Advantages and Disadvantages of Gasoline, Diesel, and LPG Fuels

Gasoline, diesel, and liquified petroleum gases—or LPG (propane and butane)—are common motor fuels used in agriculture. Write a paper that lists the advantages and disadvantages of each type of fuel. Discuss these advantages and disadvantages in the class.

Health Occupations

Common Health-Related Products from Fossil Fuels

List common products that are health related that come from petroleum products. Consider such items as bandages, adhesives, ointments, medication bottles, syringes, and so forth. Discuss alternatives to petroleum for producing each product. If no alternative can be suggested, how would the absence of this product affect people?

Artificial Body Parts

Discuss how petroleum products are used in artificial body parts. Some artificial body parts to consider are: artificial limbs, heart valves, and replacement joints. Can you suggest resources to replace petroleum products in each artificial body part?

Home Economics

Weight Management and High-Energy Foods

Use a calorie counter and plan a low-calorie diet for weight management. Also plan a high-energy diet. Compare and contrast these two diets.

Utility Spokesperson to Speak About Energy Savings in the Home

Invite a spokesperson from the electric or gas utility to speak to your class about measures to save energy in the home. Discuss cost and effectiveness of each measure.

Flammability Test of Natural Fibers and Untreated Synthetic Fibers

Obtain samples of wool, cotton, and synthetic materials that have not been treated with flame retardant. Demonstrate the flammability of synthetic fibers and compare to the flammability of natural fibers such as wool and cotton.

Industrial Technology

Changes in Engine Design Since 1973

In the library, use magazines such as *Popular Science* to research changes in automotive engine design since 1973. How does each change affect fuel economy and air quality?

Solar Water Heater

Design and build a solar water heater. Determine the feasibility of using solar energy to heat water in your area. Consider construction cost, backup supply, dependability, fuel savings, and time needed for the savings to pay back the initial cost.

HOW DO WE MEASURE THE ENERGY STORED IN A FOSSIL FUEL?

Introduction

The power plant technicians at Edison Power think of energy as the capacity to do work. A lump of high-grade coal releases more energy when burned and can do more work than the same size lump of low-grade coal. But how does a lump of coal store energy? And how is that energy released to do work?

The answers to both these questions lie in the chemical bonds you read about earlier. Compounds such as hydrocarbons store chemical energy in the bonds that hold the atoms of the compound together. The energy in the bonds is measured in units called calories. When a substance burns, some of the energy stored in its chemical bonds is released as heat. The combustion reaction for methane is shown as an example below.

$$CH_4 \ + \ 2O_2 \ \longrightarrow \ CO_2 \ + \ 2H_2O \ + \ \text{heat energy}$$

(methane) (oxygen)   (carbon   (water)
                dioxide)

Since heat is a form of energy, the amount of heat is measured in calories. The number of calories released when one gram of a compound burns is the **heat content**.

At a power plant, heat is the desirable product of the combustion of a fossil fuel. Heat is used to boil water and make steam. Technicians at Edison Power periodically determine the heat content of the lignite coal burned in the plant's furnaces. The higher the heat content of the coal, the less fuel will need to be burned in the furnace. In this lab you will use methods similar to those used at Edison Power.

Nutritionists also make use of the heat content. Like fossil fuels, food is a mixture of chemical compounds that store energy in their chemical bonds. Nutritionists determine how many calories are in a particular type of food by finding out the food's heat of combustion. In general, a gram of fat or oil stores about 2.25 times more energy than do starches or sugars. If your diet contains too much fat you will need to exercise more to burn up excess stored energy.

Purpose

In this lab, you will estimate and compare the heat content of various hydrocarbon fuels by measuring the heat absorbed by water and weighing the fuel before and after burning.

Lab Objective

When you've finished this lab, you will be able to —

- Estimate the heat content of a fuel by measuring the increase in temperature of water heated by burning the fuel.

Lab Skills

You will use these skills to complete this lab —

- Measure mass, volume, and temperature.
- Burn fuels and handle hot objects safely.

Materials and Equipment Needed

candle, canned heat, or paraffin	matches
watch glass (optional)	water
thermometer, Celsius	hot pads
ceramic fiber pad	glass rod
graduated cylinder, 100 ml	
safety goggles	
triple-beam balance (121 g @ 0.01 g)	
calorimeter kit	

Pre-Lab Discussion

Whenever liquid water is heated, its temperature is raised in a predictable way. If one gram of water absorbs one calorie of heat, the temperature of that one gram will increase by one Celsius degree. This relationship (one gram, one calorie, one degree) holds true for any amount of water at ordinary temperatures. It doesn't matter what fuel is used to supply the heat.

Safety Precautions

- Keep your hands, hair, face, and personal belongings away from the flame of the burning fuel.
- Wear your safety goggles at all times when you work in the lab to avoid serious injury to your eyes.
- Metals absorb heat. Use hot pads to handle them.

Put on your safety goggles now. Leave them on throughout the entire lab procedure.

Method

1. **Measuring Mass**

    a. Weigh the fuel and container on the triple-beam or digital balance on your lab counter.

    b. Record the mass, to the nearest 0.01 g, on Line A of Data Table 1.

2. **Measuring Volume**

    a. Look at Figure L1-1 to see how to measure the volume in a graduated cylinder.

    b. Use a graduated cylinder to measure 100 ml of water.

    c. Pour the 100 ml of water into the flask.

    d. Record the volume on Line C in Data Table 1.

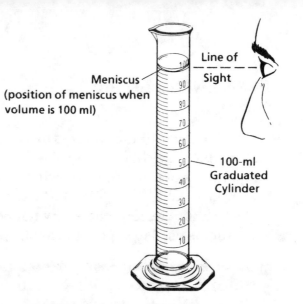

**Figure L2-1
Measuring
with a
graduated
cylinder**

Line of
Sight

Meniscus
(position of meniscus when
volume is 100 ml)

100
90
80
70
60
50
40
30
20
10

100-ml
Graduated
Cylinder

3. **Measuring Temperature**

    a. Use a thermometer to measure the temperature of the water in the flask.

    b. Record this value in °C on Line D of Data Table 2.

4. Do the following:

    a. Suspend the flask from the support plate using the plastic collar.

    b. Place the support plate on top the calorimeter can.

    c. Slide the support plate to one side to provide a small opening as a vent.

    d. Place the thermometer inside the flask.

    e. Place your fuel and container on the ceramic fiber pad. See Figure L2-2 for the completed setup.

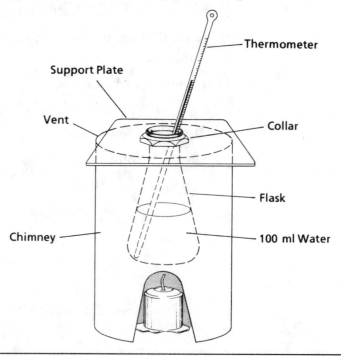

Thermometer

Support Plate

Vent

Collar

Flask

Chimney

100 ml Water

**Figure L2-2
Completed
lab setup**

Applied Biology/Chemistry

Be sure all flammable items are moved away from your lab setup.

Avoid contact with the flame!

5. Light the fuel.

6. Quickly lower the calorimeter and flask over the burning fuel.

7. Watch the thermometer level rise. Allow the water temperature to rise at least 20 C° above the starting temperature.

    Then do the following steps in rapid order.

    a. Using a hot pad, lift the calorimeter and set it aside.
    b. Blow out the flame.
    c. Gently stir the water in the flask with a glass rod.
    d. Record the final temperature of the water on Line E of Data Table 2.

Allow your unburned fuel to cool.

8. When the unburned fuel has cooled:

    a. weigh the unburned fuel and its container;

    b. record this value to the nearest 0.01 g on Line B of Data Table 1.

Observations and Data Collection

Copy the data tables below into your ABC notebook and record your results there. **DO NOT WRITE IN THIS TEXTBOOK.**

DATA TABLE 1

		Candle	Canned Heat	Paraffin	Other
A.	Starting mass of fuel and container	___	___	___	___
B.	Final mass of fuel and container	___	___	___	___
C.	Volume of water in flask	___	___	___	___

DATA TABLE 2

		Candle	Canned Heat	Paraffin	Other
D.	Starting temperature of water	___	___	___	___
E.	Final temperature of water	___	___	___	___

Calculations

Do the following calculations in your ABC notebook.

Do each of the calculations below for the fuel you used in the experiment. If you have time left, do the experiment again with a different fuel.

F. Which type of data did you record that allowed you to find the increase in the temperature of the water during heating? _____, _____. Calculate and record this temperature increase next to the fuel you used. Use the proper units for temperature.

Candle _____   Canned Heat _____

Paraffin _____   Other Fuel _____

G. Use the density of water (one g/ml) and the volume of water you measured (line C) to calculate the mass of water heated by your fuel. Record this mass next to the fuel you used. Use the proper units for mass.

Candle _____   Canned Heat _____

Paraffin _____   Other Fuel _____

H. Calculate the amount of heat used to raise the mass of water in your flask by the number of degrees you recorded in F. Record the amount of heat next to the fuel you used. Use the proper units for heat.

Candle _____   Canned Heat _____

Paraffin _____   Other Fuel _____

I. Which type of data in your table allows you to find the mass of fuel consumed during heating of the water? _____, _____. Calculate and record this mass of fuel consumed next to the fuel you used. Use the proper units for mass.

Candle _____   Canned Heat _____

Paraffin _____   Other Fuel _____

J. If heat content is the number of calories that one gram of a fuel can supply when burned, calculate the heat content of the fuel you used. Record this value next to the fuel you used. Use the proper units for heat content.

Candle _____   Canned Heat _____

Paraffin _____   Other Fuel _____

Cleanup Instructions

- Pour water down the lab sink drain.

- Clean up any loose wax drippings and discard in the trash.

- Use a paper towel to remove soot from the bottom of the flask.

- Store fuels and calorimeter for future use.

- Return the balance, thermometer, and graduated cylinder to their proper storage places.

- Store matches in a cool, dry place away from flammable materials.

WRAP-UP

Conclusions

Answer the following questions in your ABC notebook.

1. Create a new table with enough room to include both the data you recorded and the results of all calculations you just made.

2. Which of your measurements (mass, volume, or temperature) do you feel may have had the greatest error? How would such an error have affected your calculation of heat content?

3. Compare your experimental values for heat content to values in Table L2-1. Why are your values less than those below?

Table L2-1: Heat Content

Candle wax	11,300 cal/g
Paraffin	11,300 cal/g
Canned heat	4,738 cal/g
Ethanol	7,122 cal/g
Methanol	5,340 cal/g

4. Classify the natural resources consumed in this experiment as limited or unlimited and renewable or nonrenewable.

Challenge Questions and Extensions

Answer the following questions in your ABC notebook.

5. How might you verify that some of your fuel was actually converted to water vapor during combustion?

6. What mass of candle (or canned heat or paraffin) must be burned to raise the temperature of a liter of water from room temperature to the boiling point? Use heat contents from Table L2-1.

AIR AND OUR ATMOSPHERE

THINK ABOUT IT

- What physical problems might this girl be having while riding her bike on a heavily traveled road?

- Have you ever been in a place where the air was heavily polluted? How did your body respond?

- What are two ways air is being used in the above scene?

SUBUNIT OBJECTIVES

After you complete this subunit, you will be able to—

1.   Describe the composition of air.

2.   Explain the two main reasons that air is an essential natural resource.

3.   Analyze the causes and effects of problems related to air as a natural resource.

4.   Suggest ways to reduce problems related to air.

5.   Identify jobs that are concerned with air and air quality.

LEARNING PATH

To complete this subunit, you will—

1.   Read the text.

2.   Take part in class discussions and activities.

3.   Do Laboratory 3, "What Is the Normal pH of Rain?"

4.   View and discuss the video problem, "The Greenhouse Effect."

JOB PROFILE: AIR POLLUTION CONTROL TECHNICIAN

Carmen P. is an Air Pollution Control Technician. She works for a state air quality-control agency. She is primarily a field technician, so she installs, operates and sometimes repairs air-sampling equipment. She also takes air samples that are analyzed to find out if harmful gases or particles are in the air. She takes readings of wind speed, humidity and temperature because these factors affect pollution.

Carmen began her career by getting a two-year degree in chemistry. For a couple of years, she worked in the state air pollution laboratory. She says, "I really like the outdoors, and I like working with equipment and tools. I decided I would be a lot happier as a field technician. So I went back to technical school and took another year of courses in instrumentation. That prepared me for the field job." Carmen still works part-time in the lab, but much of her time is spent driving around to the different air-sampling stations, checking instrumentation and taking readings.

"Using our instrumentation, I can take readings of SO_x, NO_x, CO_2, CO and other gases. We also analyze for solid particles—particulates," says Carmen. She goes on, "A lot of the pollution in this area is related to auto emissions. Our city doesn't have a very good public transportation system. People are out there in their own cars, not thinking too much about what they put into the air."

What Is the Composition of Air?

Air is a mixture of gases. A mixture is composed of two or more substances, each of which retains its own chemical properties.

If air is put under pressure and cooled, it becomes a pale blue liquid. When air in a liquid state starts to warm up, the gases that make up the mixture boil off at different temperatures. This shows us which gases can be found in air.

Figure 3-1 shows you the composition of a typical sample of dry air at sea level. A sample of air taken from a few feet above a lake or stream may be about 3% water. Air above a desert may be near 0% moisture. Most gases in air are inorganic substances. The molecules of inorganic substances contain few or no carbon atoms.

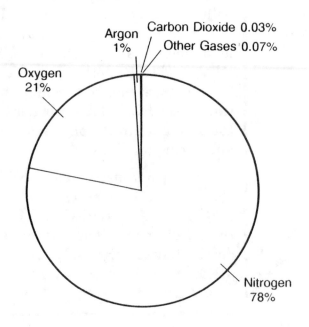

**Figure 3-1
Composition
of air at sea
level**

ACTIVITY 3-1

- Answer these questions about the graph in Figure 3-1 in your ABC notebook.
  1. What percent of the air is oxygen?
  2. What percent of the air is carbon dioxide?
  3. What gas makes up the greatest percentage of air?
  4. If the percent of carbon dioxide in the air and the percent of nitrogen in the air were both to change by one tenth of a percentage point, which change would you guess might have the greater impact on life on Earth? Why?

Air also contains particles of matter, called particulates. Particulates get into the air from various sources. Particulates include dust, pollen, ash and other wind-borne matter.

Particulates in Lignite Stack Emissions

When operators at Edison Power burn coal in the furnace, ash particulates will be produced. Some of the ash, flyash, will be drawn toward the stack. The stack is where hot gases and other products of combustion are emitted.

The Environmental Protection Agency is very strict about the amount of flyash that can escape the stack. They require power plants to capture 99% of the ash that's drawn upward. The best way to collect flyash is by using electrostatic precipitators. They're called ESPs for short.

An ESP works by collecting charged particles in the smoke. All gases and ash particles leaving the furnace pass through an electrically charged field. This field gives all the ash particles a negative charge. The ash particles then pass between large plates that have a positive charge. The negatively charged particles stick to the positively charged plates. ESPs can capture more than 99% of the ash particles before they get to the stack.

- Using research materials available to you in your school library, community library, and other sources, answer the following questions:
  1. What are some problems associated with ash particulates escaping into the air?
  2. How do they affect you and your community?
  3. What jobs have been created to combat these problems?
- Write your answers to these questions in your ABC notebook. Be prepared to discuss your findings.

Why Is Air Important as a Natural Resource?

Air is an essential natural resource because it supports life and it supports combustion.

Air to Support Life

Animals breathe in air and absorb the oxygen (O_2) in it. The O_2 is carried by the blood to cells throughout the body. Inside the cells, O_2 reacts with digested food to release energy. This energy is needed for all life processes.

ACTIVITY 3-3

Contact a respiratory therapist at a local clinic or hospital. These therapists generally work with people who have breathing problems. Get answers to these questions:

1. How does a respiratory therapist measure the amount of air that a patient can take into his or her lungs?
2. How can a respiratory therapist determine how much oxygen a patient has in his/her bloodstream?
3. Under what circumstances is pure (100%) oxygen administered to an ill person and what effects does it have?
4. Find out how air quality in the environment affects the patients treated by the respiratory therapist.

Air also contains the carbon dioxide (CO_2) that plants need in order to grow. As you can see in Figure 3-1, the air has a relatively small amount of carbon dioxide. Plants use this CO_2 to make their own food in the process of photosynthesis. Since plants are the basis of the entire food supply, you can see how important the carbon dioxide in the air is.

Nitrogen is another important gas in air. During electrical storms, nitrogen in the air is turned into substances called nitrates. These nitrates fall to the ground and become important nutrients for plants.

Air to Support Combustion

Air is needed to burn things. Specifically, the oxygen in the air is used in a combustion process. Remember that in the combustion of a fossil fuel (or any hydrocarbon), oxygen (O_2) combines with the hydrogen (H) and carbon (C) in the fuel. The products of this reaction are heat, carbon dioxide (CO_2), and water (H_2O). The nitrogen in air helps control the combustion process. It keeps the reaction between fuels and oxygen from being too violent.

Combustion is a type of chemical reaction called an oxidation reaction. One way to think of oxidation is to picture a substance combining with oxygen. Many substances undergo this kind of oxidation. When a fuel burns, it oxidizes very rapidly. Other substances oxidize more slowly. Iron is such a material. When iron oxidizes, it forms rust.

Lignite and Efficient Combustion

Lignite will not be burned as lumps of coal at Edison Power. Power plants that use coal first crush it into a powder. The lignite is pulverized so that more surface area of the coal is exposed to air, making combustion more rapid and efficient. Then the coal is blown into the furnace. This stream of pulverized coal is mixed with air. The fuel-to-air ratio is very carefully controlled. The reason for doing this is to ensure that the coal gets exactly the right amount of air for efficient burning. In other words, Edison wants to get as much energy as possible from the lignite.

What Are the Problems and Issues Related to Air?

We have to be very careful not to add harmful amounts of polluting substances to air. If we do, we may threaten air's ability to support life.

What Is Acid Rain?

When we burn fossil fuels, we produce oxides of nitrogen and sulfur. Sulfur and nitrogen oxides can have several forms, such as SO_2, SO_3, NO, or NO_2. The formulas SO_X (sulfur oxides) and NO_X (nitrogen oxides) are sometimes used. (X can represent any number from one to three.) These formulas are used to designate sulfur and nitrogen oxides in general. They don't identify a specific sulfur or nitrogen oxide.

Controlling Air Pollution in Mexico City

Mexico City is the world's largest city. With over 30,000 industries in the area and an estimated 2 million cars on the road every day, the city has developed serious pollution problems. To make matters worse, the city is located in a valley surrounded by mountains. In the cooler times of the year, the pollution is forced down into the valley. People get sick easily. Almost everyone feels the effects of increased levels of carbon monoxide, sulfur dioxide, nitrogen oxides, hydrocarbons, and lead.

In 1989, the government instituted an experimental pollution control program that requires all citizens to forgo using their cars for one day a week. Anyone who does not cooperate pays a stiff fine. (The fine is approximately equivalent to one month's pay at Mexico's minimum wage.)

The result was that the number of cars was reduced by about 500,000 per day, and the pollution levels dropped 10 to 15%. People noticed the difference, and there has been a lot of popular support for continuing the program, at least during the periods of the year when pollution is at its worst.

An influential environmental group in Mexico, known as Grupo de Cien, or Group of 100, supports the program but points out that much more has to be done if Mexico City's pollution problems are to be solved. They point out that industrial polluters are not regulated strongly enough and that unleaded gasoline and catalytic convertors need to be required on all vehicles driven in Mexico.

ACTIVITY 3-5

- Discuss the following questions:

  1. Why do you think the pollution problem in Mexico City had to become so severe before anything was done about it?

  2. In your community, what causes the greatest amount of pollution—auto emissions, industrial pollution or some other source? How do you know? If you don't know, how could you find out?

  3. How would you and your family adjust if you were unable to use your car one day of the week?

When SO_X and NO_X get in the air, they combine with water in the atmosphere to form acids. For example, SO_3 (an oxide of sulfur) can combine with water (H_2O). This forms sulfuric acid (H_2SO_4). When it rains, the acids fall to earth as acid rain.

$$SO_3 + H_2O \longrightarrow H_2SO_4$$

Acid rain is a serious problem in some parts of the world, including parts of the U.S. The effects of acid rain are still being studied. Most organisms cannot survive a pH of less than 4.0. You need to be aware of acid rain because it could affect your food and water supply, as well as your health.

Acid Rain and Plants. Acid rain affects plants in two ways:

- Leaves are often damaged, reducing a plant's ability to produce its food.

- The soil becomes acidic, interfering with the plant's ability to absorb nutrients from the soil.

Depending on the geology of an area, the effects of acid rain on the soil may be neutralized. This may happen, for example, in areas with lots of limestone.

ACTIVITY 3-6

When a material is too acidic—low pH—or too basic—high pH, it can be adjusted by neutralizing the acid or base. Neutralization is a reaction between an acid and a base that produces water and a salt. The salt is made up of the negative ion from the acid and the positive ion from the base.

Now, let's look at the reaction between hydrochloric acid and sodium hydroxide.

1. Get a set of molecule cards from your teacher.

2. Find the cards that are labeled:

   - Hydrogen (need two)
   - Oxygen
   - Chlorine
   - Sodium

3. Put the two hydrogen cards and the oxygen card together to form a water molecule—H_2O.

4. Remove one of the hydrogens from the water to form a hydrogen ion—H^+—and a hydroxide ion—OH^-.

5. Place the hydrogen ion on the chlorine card to form the molecular form of hydrochloric acid—HCl.

6. Place the hydroxide ion on the sodium card to form a molecular form of sodium hydroxide—$NaOH$.

NOTE: When a strong acid or a strong base dissolves in water, the molecules break apart into ions—the molecule ionizes. For example, when hydrochloric acid or sodium hydroxide is dissolved in water, its molecules ionize.

7. Break the sodium hydroxide and hydrochloric acid molecules into ions of Na^+, Cl^-, H^+, and OH^-.

8. Put the hydrogen ion—H^+—and hydroxide ion—OH^-—together to form water—H_2O.

NOTE: When hydrogen ions and hydroxide ions get together, they form water—H_2O. The sodium and chloride ions will remain as ions as long as they are in solution, but when the water is evaporated, they will form sodium chloride or table salt.

9. Put the sodium—Na^+—and chloride—Cl^-—ions together to form the salt sodium chloride.

10. Write a chemical equation for the neutralization reaction between hydrochloric acid and sodium hydroxide.

Acid Rain and Water Environments. Acid rain also affects water environments.

- The aquatic plants that are the food for many fish are damaged.

- The water itself can become too acidic for the fish and other animal life.

Acid Rain and Human Health. Besides causing acid rain, the SO_X and NO_X in the air have direct effects on human health.

- When people breathe polluted air, the SO_X and NO_X in the air react with the moisture in the lungs. This produces acid that irritates the linings of the lungs and upper respiratory tract.

We should point out that even normal rainfall is usually slightly acidic. It's acidic because water in the air reacts with carbon dioxide (CO_2) in the air. This produces carbonic acid.

$$H_2O + CO_2 \longrightarrow H_2CO_3 \text{ (carbonic acid)}$$

The term acid rain is used only when rainfall is excessively acidic. Also, any form of precipitation can be acidic. This includes hail, sleet, and snow.

How Is Acidity Measured?

The level of acidity or alkalinity of a liquid is referred to as its pH. If the pH is lower than 7, it is acidic. If the pH is higher than 7, it's alkaline. If the pH is exactly 7, the liquid is not acidic or alkaline. Its pH is referred to as neutral. The pH of pure water is 7. The pH of a liquid can be measured with a type of paper known as indicator paper, or with an electronic meter. Figure 3-2 shows the pH of some common items on a pH scale.

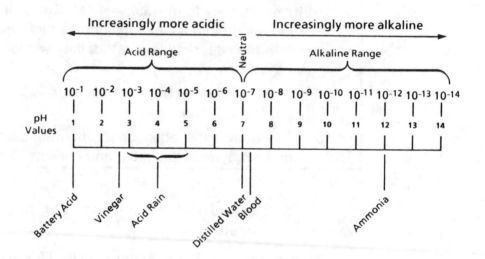

Figure 3-2 pH scale

A change of one pH unit means the acidity is ten times greater or lesser. This means that rain with a pH of 4.6 is ten times more acidic than normal rain of pH 5.6. In a body of water, a tenfold change of pH from 5.6 to 4.6 would greatly affect the life processes of aquatic organisms.

Applied Biology/Chemistry

Cathy M. is an aquatic biologist for the water commission of a state in the Northwestern U. S. She collects samples from natural water bodies, industrial wastes, or other water sources. She performs chemical and physical tests in the field and in the laboratory. Sometimes she sets up monitoring equipment to obtain information on water flow, temperature, pressure and other factors.

Part of Cathy's job is to regularly monitor the effects of acid rain by measuring pH samples of small lakes in her district of the state. She takes samples of lake water on a regular basis and tests them for acidity. She sends the results to the state water commission.

ACTIVITY 3-7

Below are two sets of readings from Cathy's logbook, taken one year apart in early May.

- In your ABC notebook, make a table for these readings.

  May '88 readings:
  Dodgers' Lake - pH 7.5, Pirates' Lake - pH 5.8, Astros' Lake - pH 6.8, Orioles' Lake - pH 4.6, Angels' Lake - pH 7.1.

  May '89 readings:
  Dodgers' Lake - pH 7.4, Pirates' Lake - pH 4.8, Astros' Lake - pH 6.9, Orioles' Lake - pH 3.8, Angels' Lake - pH 7.2.

- Examine the readings above and discuss these questions:

  1. In one lake, Cathy has seen mostly mature fish. She has seen very few young fish. What conclusions might she draw from this? At which lake would you guess that she has made this observation?

  2. Which lake would you expect to be almost without plant or animal life?

  3. In which lakes is acidity decreasing?

Carbon Monoxide and Smog

When an automobile engine burns gasoline (a fossil fuel), carbon monoxide (CO) is a product. This pollutant is in addition to the sulfur and nitrogen oxides that are produced in the combustion of fossil fuels. Carbon monoxide is produced because the combustion of gasoline in an automobile engine is not complete. Carbon monoxide in the air is dangerous to the health of animals, including people. When breathed, carbon monoxide can replace oxygen in the blood. But it cannot be used by the cells. Consequently, carbon monoxide is a poison to living things.

Carbon monoxide is also a major component of smog (Figure 3-3). In addition to carbon monoxide and oxides of sulfur and nitrogen, smog has one more major component: ozone.

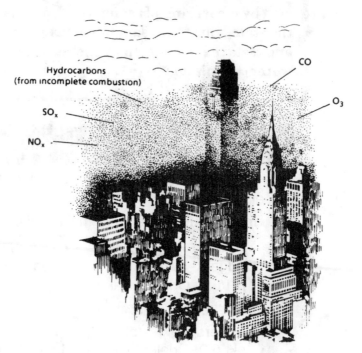

**Figure 3-3
Smog**

Ozone

Ozone (O_3) is a form of oxygen. It occurs naturally in the atmosphere. Ozone is also produced in certain reactions. Have you ever noticed the strange odor around an electric train set? What you smell is ozone. It's produced when the oxygen in the air interacts with electricity.

Ozone is also produced when sunlight activates automobile exhausts. Far too much ozone gets into the air of our cities, largely due to automobile exhausts. Ozone is very unstable and breaks down rapidly into O_2 and a lone oxygen atom. This lone oxygen atom is highly reactive. When it comes into contact with things like metals, building materials, or exposed body tissues, it will oxidize them. Ozone is very harmful to the body.

Ozone Destruction

Too much ozone near the Earth is harmful. In the outer layer of Earth's atmosphere, however, ozone is useful. It absorbs ultraviolet (UV) radiation beamed to Earth by the sun. This absorption limits the amount of UV able to reach the Earth's surface. In this way ozone acts to protect you from the harmful effects of UV. These harmful effects include skin cancer and cataracts.

Research has shown that ozone is destroyed by commercial compounds called chlorofluorocarbons. Chlorofluorocarbons are like hydrocarbons but contain highly reactive chlorine (Cl) and fluorine (F) atoms intead of hydrogen. They are used as coolants in air conditioners and refrigeration systems. They're also used as propellants (gases that push out the contents of the can) in aerosol spray cans. Defective air conditioners and refrigeration systems can give off large amounts of chlorofluorocarbon gases. These gases rise into the upper atmosphere to react with ozone and destroy the Earth's protective UV shield. Leaks in automobile air conditioners are a major way that chlorofluorocarbons get into the air.

Freon and the Ozone Layer

Charles De Mann is an automotive mechanic and garage owner. He employs two other mechanics. He makes a reasonable middle-class living from the garage.

About ten percent of De Mann's automotive business is air-conditioning repair. When he repairs air-conditioning systems, he usually has to release the used freon into the atmosphere. Charles knows that freon is one of the gases that interacts with ozone. He doesn't like the idea of letting the freon out into the air.

"There's a device on the market that would allow the used freon to be reused. It works like this," explains Charles. "You just hook it into the car's air conditioner. Then you pump out the old freon into a holding tank. Inside the tank, the water and other impurities are filtered out. Then you put the clean freon back into the air conditioner. You don't throw out the old freon. You clean it up and reuse it."

Charles would like to buy such a device, but the price tag is $3000. "Putting in new freon costs only about $8 to $10 per car. It would take me about ten years to recover the cost of a freon cleaner," says Charles. "The Environmental Protection Agency may require garage owners to use it someday. If they do, I might just have to stop doing air-conditioning work. $3000 is a lot of money to me."

There are thousands of mechanics like Charles all over the country.

ACTIVITY 3-8

- Discuss these questions as a class:
    1. What could be done to encourage auto mechanics like Charles to buy and use freon filtering systems and to stop putting used freon into the atmosphere?
    2. Should Charles pass the cost of the cleaner on to his customers?
    3. How might that affect his business?
    4. Would you be willing to pay more for air-conditioning repair in order to ensure that the used freon is not released into the air?

Carbon Dioxide and Global Warming

Carbon dioxide is another product of the combustion of fossil fuels. This product, too, may present a problem. Plants need carbon dioxide from the air to make their own food. It's possible, however, for the atmosphere to have too much carbon dioxide. And today, the huge quantities of CO_2 produced by the burning of fossil fuels may be too much.

It is speculated that the excess amounts of CO_2 in Earth's atmosphere act like a blanket (Figure 3-4). When sunlight passes through the atmosphere and strikes the ground, it turns into heat. The blanket of CO_2 traps this heat close to the earth. This condition is known as the greenhouse effect. The blanket of CO_2 traps heat like glass traps heat in a greenhouse.

In moderation, the greenhouse effect helps makes Earth a livable planet for plants and animals. Without it, Earth would be too cold for most familiar life forms. The greenhouse effect becomes a problem only when the blanket of CO_2 gets too thick. If the blanket of CO_2 becomes too thick, it may hold too much heat next to the Earth. This condition may cause temperatures to increase around the world.

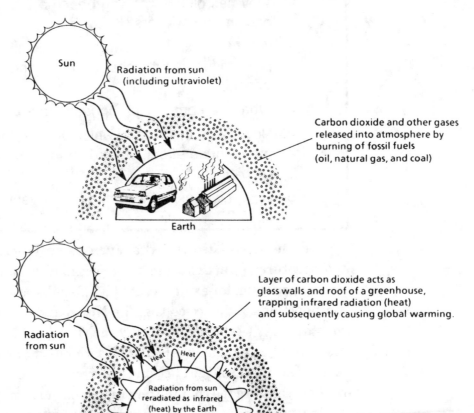

Figure 3-4 The greenhouse effect

ACTIVITY 3-9

- Use the *Reader's Guide to Periodical Literature* in the school library or the local library to look up information on global warming.

  The *Reader's Guide* is organized by yearly volumes; for the current year, monthly issues are available. Each volume is arranged alphabetically by subject, title or author. Your can look up "global warming" or "greenhouse effect" under the subject section of the guide.

  Some magazines that may have good, easy-to-read articles on global warming include *Science News*, *Time*, *Newsweek*, *Science*, and *Technology Review*.

- Answer these questions in your ABC notebook and prepare to share them with the class.
    1. Why has global warming become a potential problem only in recent years?
    2. What tools and procedures do scientists use to make predictions about global warming and the greenhouse effect?
    3. What are some major consequences of global warming?
    4. How might the area in which you live be affected by the greenhouse effect, according to some scientific predictions?
    5. How might global warming be prevented or slowed down?

Some scientists think that even a small increase in the temperature of Earth's atmosphere could have disastrous results. Some speculate, for example, that a 1.3-Celsius-degree increase could melt portions of the ice caps at the North and South Poles. Some coastal cities would be flooded by the rising sea level. Agriculture also would be seriously affected by a global change in climate. Some fertile farmlands might become deserts. On the positive side some regions that are too cold in the present climate might become major agricultural producers if Earth's atmosphere warmed up.

ACTIVITY 3-10

One of the causes of excess CO_2 in the atmosphere is the burning of gasoline in automobile engines.

- In your ABC notebook, keep a record of how your car, or your family's car, is used each day for a week. Record each trip made and the mileage involved.

- Compare your record with that of other students in your class.

- Analyze all of the records to see what activities involve the greatest car use.

- Discuss these questions:
  1. Which activities, if any, can be accomplished without a car?
  2. Which activities, if any, would you be willing to carry out without a car?

What Might Be Done About Problems Related to Air?

Technology, the application of scientific principles, has given us a number of ways to help keep our air cleaner. Power plants and other industries that burn fossil fuels use devices called scrubbers. Scrubbers remove sulfur oxides from waste gases. Power plants also use electrostatic precipitators to keep coal ash from getting into the air.

Improvements to car engines are resulting in a higher percent of fuel being turned to energy. Emission gas recyclers (EGRs) are now installed on new cars. EGRs route auto exhaust back to the engine for the combustion of unburned hydrocarbons. Catalytic convertors help control emissions of nitrogen oxides.

Aerosol products have been changed to eliminate the use of chlorofluorocarbons in many products. In 1989, representatives of eighty-six nations resolved to ban chlorofluorocarbon production by the end of the century. However, carrying out the resolution will be costly and complicated. In the meantime, we can keep air conditioners and refrigeration systems in good repair. This will prevent some chlorofluorocarbons from escaping into the atmosphere.

Can you think of other ways to keep air clean?

Looking Back

Air is a mixture of gases. **Nitrogen** makes up about 78% of the air you breathe. **Oxygen** makes up about 21% of it. The remaining 1% is made up of **carbon dioxide**, **argon** and other gases. Air is such an important resource because it supports life and is used in the combustion process. The air's ability to support life is threatened by harmful amounts of polluting substances. **Acid rain**, **smog**, and **ozone destruction** result from pollution. These problems affect other resources as well as air. Technology has given us a number of ways to help keep our air cleaner, including electrostatic precipitators in power plants and emission controls in cars.

Vocabulary

The words and phrases below are important to understanding and applying the principles and concepts in this subunit. If you don't know some of them, find them in the text and review what they mean. They're listed in the order in which they appear in the subunit.

particulates	acidic
carbon dioxide	alkaline
oxidation	carbon monoxide
sulfur oxides (SO_x)	smog
nitrogen oxides (NO_x)	ozone
acid rain	chlorofluorocarbons
pH	greenhouse effect

Further Discussion

- What is the quality of air in your community? What factors contribute to this quality?

- Find three newspaper or magazine articles related to problems with air in our society. Discuss them in class and devise possible solutions to the problems in the articles. As a class, you may prepare a bulletin board or other display for your school or classroom to create awareness of the problems concerning air pollution.

- You work in a pre-school located in a large city with serious air pollution problems. On some days, the pollution is so bad that the children can't go outside to play. Working in small groups, write a note to the parents. In the note, explain the health risks that make it necesssary to keep the children inside. Write another explanation that would help the children understand what causes air pollution and why they can't play outside. Use terms children will understand. Have a representative from your group read the explanations in class.

Activities by Occupational Area

General

Acid Rain Speaker

Invite someone from the Soil Conservation Service or the State Wildlife Department to speak to your class on the impact of acid rain. Try to identify local sources of sulfur and nitrogen oxides that can cause acid rain.

Model of Atmosphere

Suspend plastic beads (small colored beads from a hobby shop) in a quart jar of cooking oil. Use a different color to represent each of the gases that make up the atmosphere. Weigh the beads to get the right proportion of each gas.

Agriculture and Agribusiness

Effect of Acid Rain on Future Food Supply

Write a paper on the effect acid rain may have on future food supply. Consider solutions to food production problems. What can be done to avoid these problems? Discuss in class the problems and solutions that you have discovered.

Effect of Global Warming on Future Food Supply

Discuss the effect global warming from the greenhouse effect would have on food production. How would the geographical areas used for food production change? What climate changes are likely to occur in geographic areas that are now used for food production?

Health Occupations

Filter Cigarette Smoke

Filter cigarette smoke through a filter paper to demonstrate some of the "dirt" that can be removed from cigarette smoke. Discuss the health hazards of cigarette smoking and inhaling "second-hand smoke."

Air Pollution Speaker from American Lung Association

Invite a speaker from the American Lung Association to discuss the relationship between air quality and health.

Home Economics

pH of Various Household Products

Collect several household products such as: shampoo, dishwashing detergent, household ammonia, glass cleaner, liquid bleach, etc. Determine the pH of each product.

Comparison of Different Sunscreens

Discuss the differences in several sunscreen products. Include in your discussion the need for sunscreens (premature aging of the skin, skin cancer, sunburn, etc.) Mention environmental changes that make sunscreen products even more important.

Comparison of Aerosol and Pump Sprays

Compare the spray pattern, mist characteristics, convenience, and economy of aerosol products and pump spray products. Discuss the environmental effects of each type of product. Compare the advantages and disadvantages of these products.

Industrial Technology

Filter Automobile Exhaust

Caution: The exhaust pipe will be hot if the engine has been running for a few minutes. Be careful not to burn your hands. Hold a filter paper firmly across the exhaust pipe opening of an automobile while the engine is running. After about five minutes, examine the particulate matter that is trapped in the filter paper. Discuss the effect of particulate matter in the air on the environment.

Compare Different Respirators for Working Around Toxic Fumes

Find out what respirators are used to work around different toxic fumes. Discuss how each type of respirator works.

WHAT IS THE NORMAL pH OF RAIN?

Introduction

As a fisheries technician you have been asked by your State Director to determine if there is an acid rain problem in your area. You know that acid rain is a problem in many areas of the country, but you need to find out what causes it. Your research shows that the oxides of nitrogen and sulfur polluting the air dissolve in rainwater and lower the pH to damaging levels. These pollutants are not the only gases that lower the pH when they dissolve in water.

Carbon dioxide is an important component of the atmosphere. When carbon dioxide dissolves in water, it forms carbonic acid according to the reaction

$$CO_2 \; + \; H_2O \longrightarrow H_2CO_3$$

The carbon dioxide in the atmosphere dissolves in raindrops as they form. The dissolved carbon dioxide in rain from unpolluted air causes the pH to be acidic.

Purpose

In this lab, you will determine the pH of normal rain and compare this pH to the pH of a freshly boiled distilled water sample and a rain sample from your area.

Lab Objectives

When you've done this lab, you will be able to —

- Compare the pH of two or more samples.
- Test for the presence of carbonic acid in distilled water.

Lab Skills

You will use these skills to complete this lab —

- Measure the pH of a solution with a pH meter or pH paper.
- Measure a liquid volume with a graduated cylinder.

Materials and Equipment Needed

pH meter or pH paper	drying tube
squeeze bulb	crushed limestone ($CaCO_3$)
pinch clamp	plastic tee
beakers (two 250 ml)	rubber tubing
small cotton plugs (2)	glass tubing
distilled water and rain samples (100 ml of each)	graduated cylinder (100 ml)
	rubber stopper—1 hole
safety goggles	lab apron

LAB PROCEDURE

Pre-Lab Discussion

Your plan is to measure and compare the pH of freshly boiled (boiling drives off any gases that are dissolved) distilled water and aerated distilled water. The pH of the aerated distilled water is the same as the pH of rain in unpolluted air. Then, you compare the pH of a sample of rainwater to the pH of the aerated distilled water. This comparison indicates if your area has an acid rain problem.

Safety Precautions

- Use care when inserting glass tubing through holes in rubber stoppers. Careless handling of glass can break the glass and cause deep puncture wounds or cuts.

Put on your safety goggles and lab apron now. Leave them on throughout the entire lab procedure.

Method

1. Get a 100-ml sample of freshly boiled distilled water from your teacher.

2. Pour the sample into a 250-ml beaker labeled "distilled water."

3. Measure the pH of the "distilled water" sample with a pH meter or pH paper.

4. Record this pH on Line A of the Data Table.

Use caution when inserting glass tubing into holes in rubber stoppers. Insert the glass tubing carefully into the hole. Carelessness can cause deep cuts and puncture wounds.

5. Assemble the apparatus as shown in Figure L3-1.

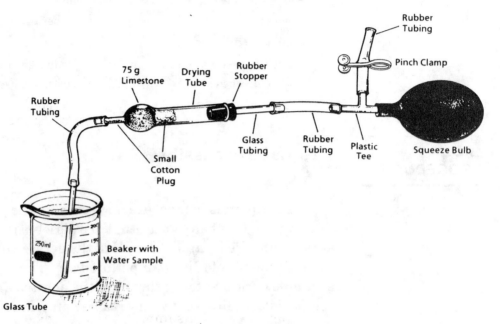

**Figure L3-1
Aeration
apparatus**

6. Aerate the "distilled water" sample as follows

    *   Submerge the end of the glass tube to near the bottom of the sample beaker.

    *   Squeeze the bulb to empty the air from the squeeze bulb.

    *   Without releasing the bulb, release the pinch clamp.

    *   Release the bulb.

    *   Replace the pinch clamp.

    *   Repeat this process twenty-five times.

7. Measure the pH of the aerated "distilled water" with a pH meter or pH paper.

8. Repeat Steps 6 and 7 until there is no change in the pH.

9. Record the final pH on Line B of the Data Table.

10. Get a 100-ml sample of rain from your teacher.

11. Pour the sample of rain into a 250-ml beaker labeled "Rain."

12. Measure the pH of the "rain" sample with a pH meter or pH paper.

13. Record the pH of the "rain" sample on Line C of the Data Table.

Observations and Data Collection

*Copy the data tables below into your ABC notebook and record your results there. **DO NOT WRITE IN THIS TEXTBOOK.***

DATA TABLE

A. pH of freshly boiled "distilled water" sample   ___

B. pH of aerated "distilled water" sample   ___

C. pH of "rain" sample   ___

Cleanup Instructions

- Disassemble the apparatus.
- Empty the crushed limestone into the trash.
- Pour the water samples into the sink.
- Wash and dry all the glassware.
- Put the glassware, rubber tubing, and pinch clamps in the proper storage area.

WRAP-UP

Conclusions

Answer the following questions in your ABC notebook.

1. What accounts for the difference in pH between the boiled water and the aerated water?

2. How many times did the aeration step have to be repeated before there was no change in the pH?

3. Why did the pH stop changing?

---

4.   How does the pH of the rain sample compare to the aerated distilled water sample?

5.   Based on your data, do you think there is an acid rain problem in your area?

Challenge Question and Extensions

Answer the following questions in your ABC notebook.

6.   If you have time, get a sample of a freshly opened carbonated soft drink and measure its pH.

7.   How does the pH of the carbonated soft drink compare to the sample of aerated distilled water?

8.   What might cause the difference in pH between the soft drink and the aerated water?

WATER AS A NATURAL RESOURCE

THINK ABOUT IT

- Do you use water for recreation? If so, how?

- Other than recreation, what are the most important ways you use water?

- How would your life be different if suddenly you could use only two gallons of water a day?

SUBUNIT OBJECTIVES

After you complete this subunit, you will be able to —

1. Explain how the water cycle works.

2. Describe what an aquifer is and how it is maintained.

3. Relate three physical and chemical properties of water to its importance as a natural resource.

4. Explain ways that physical and chemical changes in water may affect its quality.

5. Explain factors that influence the amount of water available for use.

6. Propose steps that industries and communities may take to preserve water quality and reduce water shortages.

7. Assess ways that jobs are affected by the quality and quantity of water.

LEARNING PATH

To complete this subunit, you will —

1. Read the text through "What Are the Sources of Water?"

2. Take part in class discussions and activities.

3. Do Laboratory 4, "How Do Plants Control Water Loss?"

4. Read the text through "What Problems and Issues Are Related to Water?"

5. Take part in class discussions and activities.

6. Do Laboratory 5, "How Does Acid Rain Affect Water Quality?"

7. Read remainder of the subunit.

8. Take part in class discussions and activities.

9. Do Laboratory 6, "How Efficient Is a Water Filter?"

10. View and discuss the video problem, "Water Rights, Water Wrongs."

Water as a Natural Resource

Edison Power's Water Supply

For its water supply, Edison Power will take water from Lake Edison. Water will be released from the lake as a single stream. This stream flows into the Crystal River.

ACTIVITY 4-1

- Look at the power generation diagram in Figure 1-2 in Subunit 1. How is water used in the production of electricity?

- Look at the map in Figure 1-1. What things shown on the map involve water?

- Develop a list in your ABC notebook of ways that Edison Power may affect either the quantity or the quality of the water in its environment. For example, how might the Crystal River be affected?

- For each effect Edison Power has on water, list actions that the company can take to protect water as a natural resource.

- Create a table in your notebook to show the causes, effects, and recommended actions to correct or prevent the problem.

Water (H_2O) is an extremely common substance. It covers three-fourths of the Earth's surface and accounts for 60-70% of the weight of the living world. It is also the physical environment in which many plants and animals live.

Water is something you probably take for granted, but it is essential for life. A person will die much faster from a lack of water than from a lack of food.

ACTIVITY 4-2

- Make a list in your ABC notebook of the ways you use water.
- Compare your list with lists made by other students. See who can think of the most uses.
- As a class, make a master list. Are you surprised by all the ways that you use water?
- Save this list for use in a later activity.

What Are the Sources of Water?

You are familiar with water as rain, snow, sleet, and hail. These are forms of precipitation. You are also familiar with water in oceans, lakes, rivers, and streams. These bodies of water are called surface waters. Some water is also stored below ground.

The Water Cycle

An easy way to understand water as a resource is to look at how it changes form as it moves through the environment. The movement of water through the environment is known as the water cycle (Figure 4-1).

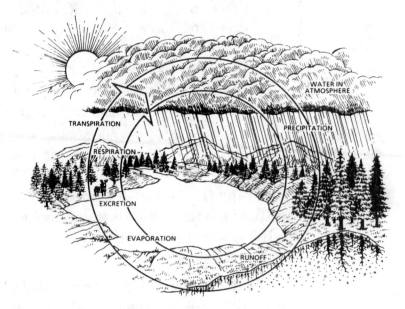

**Figure 4-1
Water cycle**

Water falls to Earth as precipitation. Then, it runs off into surface waters or seeps into the ground. Along the way it may be used by plants and animals.

Water can be returned to the atmosphere in several ways.

- Water constantly evaporates from lakes, rivers, streams, oceans, and the soil.

- The water in bodily wastes (excretion) from animals also evaporates.

- Both plants and animals release water vapor directly into the atmosphere. Animals exhale water vapor when they breathe. Plants release water vapor through their leaves in a process called transpiration.

In these ways, water on Earth is returned to the atmosphere. Once again in the atmosphere, the water may condense and fall to Earth as precipitation. The water cycle continues.

Since it covers three-fourths of the surface of Earth, water may be regarded as a plentiful resource. Water shortages, however, do occur. Even so, water doesn't disappear. It's all still there as part of the water cycle. Earth's total supply of water is never decreased.

ACTIVITY 4-3

In terms of its abundance, how should water be classified? Is it a limited resource? Is it an unlimited resource?

- Debate this issue with a classmate. Take a position that water is either limited or unlimited. Your classmate can defend the other position.

- Find as many reasons as you can to support your position.

- How do the other students in your class feel about the issue?

Underground Sources of Water

Some of the water in the water cycle is always found underground. It is stored in spaces within rocks. This area where water is stored is called an aquifer (Figure 4-2). The water in an aquifer can be used up unless water is put back. How do you think water gets into an aquifer?

Aquifers are filled when rainfall soaks into the ground and into the aquifer. The process by which an aquifer is replenished is called recharging. The area of ground through which rain soaks to refill an aquifer is called its recharge area.

Aquifers may extend for hundreds of square miles. They may supply water for cities as well as farmland. Water is pumped from aquifers for irrigation of crops, and for household and industrial use. Many country homes have wells that pump water out of aquifers deep below ground.

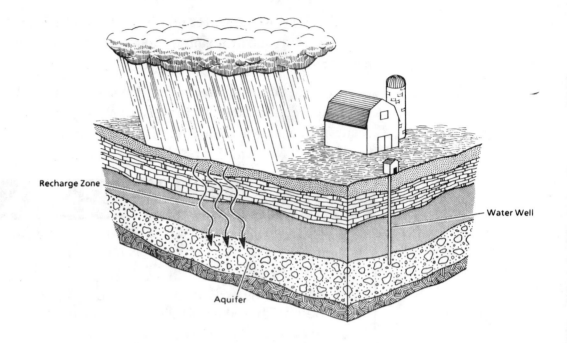

Recharge Zone

Water Well

Aquifer

Figure 4-2
Aquifer

Physical and Chemical Properties of Water

Water is a very simple chemical compound. It is made of just two elements, oxygen and hydrogen.

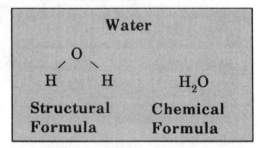

Although water has a simple chemical structure (shown above) its properties taken together are very special. Three of its most useful properties are—

1. it changes states of matter readily;
2. it is an excellent solvent;
3. it has a high specific heat.

Changes in State of Matter

The word "water" usually brings to mind a liquid. But water can also be a solid (ice) or a gas (water vapor or steam).

At 0°C, liquid water freezes to become ice. Ice itself is a valuable resource. It is commonly used to cool drinks and to keep stored food from spoiling. Ice also is used to keep blood and other animal products fresh when they are shipped.

At 100°C, liquid water boils, making steam. Steam can be used to do work. For example, steam is used to turn turbines in generating electricity.

Solvent

Water can dissolve many substances. This means water is a good solvent. Water can dissolve nutrients in the soil, minerals in rock, and even gases in the atmosphere. Because of this ability water is sometimes called the universal solvent.

High Specific Heat

When heat energy is transferred to or removed from water, the temperature of water rises or falls very slowly. The amount of heat a substance gains or loses as its temperature changes is called its specific heat. Water's specific heat is higher than that of nearly all other substances. This means that liquid water can absorb a lot of heat before it boils. This makes water a good cooling fluid.

High specific heat also makes water a temperature buffer for organisms. Plants and animals may experience wild variations in the temperature of their surroundings. Since their tissues are about three-fourths water, though, they experience only mild changes in internal temperature. Water is a good temperature buffer for living things.

- Look again at the list you made in Activity 4-2 of all the ways in which you use water.

- Indicate the state of matter—steam, water or ice—in which water is used for each item on your list.

- Indicate whether or not water is used as a solvent for each item on your list.

- Indicate whether or not water is used as a coolant or a temperature buffer for each item on your list.

- As a class, make a chart that indicates the information above next to each use of water on the list.

What Problems and Issues Are Related to Water?

When water is used as a resource, both its quality and quantity can be affected.

Water Quality

When we say that water is dirty we usually mean that we see materials suspended in it. Water quality, however, involves more than just how clear the water is. To more fully assess water quality, you have to consider substances dissolved in it, its pH, and its temperature.

JOB PROFILE: MUNICIPAL WASTEWATER TREATMENT TECHNICIAN

Suzanne D. is a wastewater treatment technician who works in the laboratory of a large treatment plant. She spends much of her time conducting lab tests. She also has to report on water quality to the federal Environmental Protection Agency (EPA).

Suzanne takes samples of effluent (water that is going out of the plant after treatment) and tests it. She also samples water at different stages of the treatment process to make sure that the process is working as it should. Suzanne runs tests to find out what suspended solid matter is in water; she also looks at water under the microscope to monitor the growth of microscopic organisms. She checks the pH of water and checks for certain chemicals such as ammonia and nitrates.

Most of the time, Suzanne works alone in her lab, but she stresses that it's important for her to have a good relationship with the plant operators. If her tests show something is wrong in the plant, she is usually the one to deliver the "bad news" to the operator. "I try to have an attitude of cooperation," she says. "We work together until the problem is solved."

Suzanne likes her job very much. She sums up, "My job may not be glamorous, but it's very important."

Suspended Matter. Suspended matter mainly affects the clarity of water. It may make water look dirty, but it often can be easily filtered out. The suspended matter may include particles of dirt, organic matter from plants and animals, or any number of other substances. Lab technicians at a water treatment plant measure the amount of suspended matter in a sample of water. This measurement is called a turbidity test.

Dissolved Substances. After it rains, the water runs off into surface waters or seeps into the ground. This allows many substances to dissolve in water. These substances may include minerals in rocks and fertilizers and herbicides from lawns and fields. Waste substances that industries discharge may dissolve in oceans, rivers, and lakes. Water in aquifers contain dissolved minerals from surrounding rocks.

Substances dissolved in water usually lower its quality. In high enough amounts, many dissolved substances can be toxic to some forms of plant or animal life.

Edison's Cooling Water

Edison Power will use large amounts of water for cooling the steam and condensing it back to water. The cooling water will be taken from the lake. Cooling water will flow through tubes in the condenser. The steam will pass from the turbine over the tubes and condense on the outside of the tubes into water as it is cooled. This same process occurs when droplets of water form on the outside of a cold can of soft drink on a humid day.

While the cooling water doesn't need to be as pure as the water used in the boiler, its quality is a concern to the company. Dissolved impurities in the cooling water are a big problem to the power plant. These dissolved impurities are commonly found in surface waters and include compounds like calcium carbonate ($CaCO_3$) and calcium sulfate ($CaSO_4$). These compounds are less soluble in warm water than in cold. At the inner surface of the tube, which is heated by the steam, the compounds will come out of solution. That is, instead of remaining dissolved in the water, they will form a solid material. This solid material will settle out and form a scale on the inside of the tubes. As this layer of scale builds up, it acts as insulation.

ACTIVITY 4-5

Answer the following questions. Then conduct the simple experiment that follows to test your answers.

1. How do you think layers of scale will affect condenser efficiency?

2. How do you think scaling might be prevented?

3. Do you think using a municipal drinking-water supply for cooling water would help scaling problems?

- Try this: Boil several cups of water in a clean pan. Continue to boil the water until it has all boiled away.

- Remove the pan from the heat source as soon as all of the water is gone.

- Examine the pan. What do you see? Does this experiment cause you to change your answer about using tap water for cooling?

pH. In the subunit on air, you learned how acid rain lowers the pH of lakes, rivers, and streams. The pH also can be lowered by runoff. For example, when compounds containing sulfur or nitrogen dissolve in runoff, the pH of the runoff is lowered. This affects the pH of surface waters that receive the runoff.

Temperature. As water boils, you probably have watched bubbles rising to the surface. These bubbles are water vapor and oxygen. When water is heated, it loses its ability to dissolve oxygen. The oxygen bubbles to the surface and escapes into the air. This means that as the temperature of a body of water increases, less oxygen will be available to support the respiration of aquatic plants and animals.

Organisms have little tolerance for great changes in temperature of their habitat. Abnormally high temperatures can kill most aquatic plants and animals in only a few hours, or even minutes.

Water Quantity

Water Quantity and Politics

Back in the early part of the century, around 1905, the people of Los Angeles were about to vote in a bond election. The bonds were slated to fund an aqueduct (a constructed waterway) to bring water south to the city of Los Angeles from a point on the Owens River, 233 miles away.

The Los Angeles Times campaigned hard for the bonds. "With this water problem out of the way, the growth of Los Angeles will leap forward as never before," the Times told its readers. The newspaper used strong language, saying that any citizen who voted against the aqueduct would be "in the attitude of an enemy of the city." It was true that the little Los Angeles River was able to provide water to a population of only two or three hundred thousand, but the whole story was never told in the newspapers.

Behind the scenes, the owner of the Times, Harrison Gray Otis, and his son-in-law, Harry Chandler, were buying up land along the route of the proposed aqueduct. In the meantime, they diverted water from certain lands to create false drought. This, in turn, kept the pressure on to bring in water from the Owens River.

In the end, the bonds passed, the aqueduct was built, and Los Angeles became the huge city that it is today.

ACTIVITY 4-6

- Discuss these questions as a class.

  1. Was there anything wrong with Harrison Gray Otis and Harry Chandler making a lot of money when the aqueduct was built? Why or why not?

  2. Do you think the voters of Los Angeles would have voted differently if the newspaper had not supported the bonds so strongly?

  3. Would they have voted differently if they had known that Otis and Chandler had a financial stake in the aqueduct?

  4. How can you decide if what you read in the newspaper is true?

One of the chief factors affecting the supply of water is rainfall. Water shortages are most severe in deserts, which may get only a few inches of rain a year.

Another reason for water shortages is that aquifers may be drained by overuse. Farmers, for example, may use millions of gallons of water a day to irrigate their crops. A large city that depends upon the same aquifer may use an equal or greater amount of water. If the demand upon the aquifer becomes too great, it may be drained faster than it's recharged.

When cities along a coast draw too much water from an aquifer, ocean water may seep in and fill it. In this case, the aquifer's value as a source of water is lost because it becomes salty.

When water is pumped out of shallow sand aquifers more rapidly than it can be replenished, the aquifers can collapse—and they usually do. This causes a sinking of the land. This has occurred around Houston and Los Angeles. Certain communities around Houston have sunk below sea level. Industries in the area used water from the aquifer instead of transporting surface water (which is more expensive). The aquifer was depleted, and it collapsed.

In large cities, there is great demand for water by homes and industries. After water is used, it must be temporarily removed from the water cycle so it can be repurified at wastewater treatment plants. This loss of water is further aggravated by urban development. Large areas of land with natural vegetation are replaced with large stretches of concrete and asphalt. Rainfall on highways, parking lots, and sidewalks evaporates instead of entering the soil or running off into streams or rivers.

What Might Be Done About Problems Related to Water?

Because water is a vital resource we should preserve its quality and make sure enough of it is around for the future.

Treating Water

As you have seen, we no longer depend upon the water cycle alone to purify our water. We build treatment plants to make water clean and safe.

The simplest way to clean water is to filter out the suspended solids. But this alone doesn't make the water safe to drink. The water still may contain living microorganisms, such as bacteria and protozoans. Many of these microorganisms can cause diseases. To kill them, water treatment plants add chlorine to drinking water. Chlorine can kill most harmful microorganisms in water.

When chlorine reacts with organic substances dissolved in the water, harmful compounds may be formed. Unfortunately, many cities can't afford the special filters used to remove some types of dissolved organic compounds or the chlorine compounds that are formed during treatment. On a small scale, however, you can do it in your home. You can use a filter with activated charcoal on your kitchen faucet. A charcoal filter removes most dissolved organic compounds.

Water Quality and Edison Power

Water quality will be an important issue for Edison Power. It needs to keep Lake Edison and Big Creek clean for the use of other industries and the residents of nearby towns.

Look at the map in Figure 1-1 in Subunit 1. Locate the water treatment unit. It will remove suspended solids and kill microorganisms in the wastewater produced by power plant activities.

ACTIVITY 4-7

Edison Power will construct six holding ponds. Locate the ponds on the map in Figure 1-1. The purpose of each pond is labeled.

- Consider each of the four factors affecting water quality:
  - suspended matter
  - dissolved substances
  - pH
  - temperature
- Explain how the holding ponds might affect any of these factors.

Reducing Water Shortages

Water shortages can be dealt with in several ways. A dam can be built across a river to create a reservoir. Water in the reservoir then can be used like water from any other source.

Another way to deal with water shortages is to divert water from an area of high rainfall to an area of need. A large river can be diverted through canals or concrete pipes to large desert cities. Water from the Rocky Mountains now supplies the needs of much of the Southern California desert.

Developers in cities can help to maintain the water supply. They can do this by leaving existing vegetation when they build new neighborhoods and shopping malls. Instead of evaporating from concrete surfaces or going into runoff, water can enter the soil. Plants absorb some of this water and store it in their own tissues.

In desert areas, landscaping should be done with plants that are native to deserts. Desert plants are adapted to dry conditions and don't require watering.

ACTIVITY 4-8

When an area is threatened by a water shortage, everyone may be called upon to conserve water. The actions of each individual citizen can be very important.

Every summer for the last three years, there has been a water shortage in the city of Marshall. The city water department has issued a conservation chart as a guideline for outside water use: watering lawns and gardens, washing cars, etc. Referring to the chart below, your teacher will assign you a zip code. Use the last digit of your zip code and the calendar below to determine which days you may water. Figure out which days you can use water for outside uses.

Key:	∇	✕	#	★	●
	0 or 9	1 or 8	2 or 7	3 or 6	4 or 5

JULY/JULIO

Sunday/ Domingo	Monday/ Lunes	Tuesday/ Martes	Wednesday/ Miercoles	Thursday/ Jueves	Friday/ Viernes	Saturday/ Sabado
						1 ∇
2 ✕	3 #	4 ★	5 ●	6 ∇	7 ✕	8 #
9 ★	10 ●	11 ∇	12 ✕	13 #	14 ★	15 ●
16 ∇	17 ✕	18 #	19 ★	20 ●	21 ∇	22 ✕
23 #	24 ★	25 ●	26 ∇	27 ✕	28 #	29 ★
30 ●	31 ∇					

- List some ways that you and your family could use less water. Decide which things would save the most water. Which would cause the most inconvenience? Which the least?

- Are any of these things worth doing even if you don't have a water shortage in your community? Why?

Looking Back

We can't live without water. Water covers three-fourths of the Earth's surface and accounts for 60-80% of the weight of the living world.

Water moves through a water cycle: it falls as rain or other precipitation, runs off into surface water, or seeps into the ground, is taken up by plants, and returns eventually to the atmosphere through the process of **evaporation** or **transpiration**. Water underground collects in **aquifers**, which are recharged by rainfall.

Water changes states of matter readily (liquid, solid, or gas) and is an excellent solvent. Additionally, water can absorb a lot of heat.

Substances dissolved in water usually lower its quality. In high enough amounts, many dissolved substances can be toxic to plant and animal life. These substances may include minerals, fertilizers, herbicides, and industrial waste.

Water quantity is directly influenced by rainfall. Aquifers may be drained by overuse, and may collapse as a consequence. The ways water is used can result in its having to be repurified at wastewater treatment plants. Because water is a vital resource, we should preserve its quality and make sure enough of it is around for the future.

Vocabulary

The words and phrases below are important to understanding and applying the principles and concepts in this subunit. If you don't know some of them, find them in the text and review what they mean. They're listed in the order in which they appear.

precipitation	states of matter
water cycle	solvent
runoff	specific heat
evaporation	temperature buffer
transpiration	pH
aquifer	suspended matter
recharge area	dissolved substances

Further Discussion

- Form a work group that includes several students. Make a list of human activities in your community that affect water quality. Discuss how these activities change water quality. Share your findings with the class.

- Divide into groups of six and play the game "Water Play." Your teacher will have the materials for the game. If there is time, change the playing cards to show local occupations. The eight characteristics of the local occupations should result in usage that is similar to the sample occupations.

- Go through the Yellow Pages of your phone book and identify 10 businesses that are dependent on water. Which ones play a role in preserving the water quality of your community?

Activities by Occupational Area

General

Hydrologic Map of Local Area

Get a hydrologic map from the county extension service. Use this map to identify the sources of water in your area. Try to identify possible sources of contamination of this water supply. What type of contamination is likely from each source?

Compare the Properties of Water Samples from Different Sources

Collect water samples from several sources such as rainwater, well water, city water, pond water, stream water, lake water, and wastewater at different stages of treatment. Visually compare the samples. Compare the odors of each sample. Measure the pH of each sample. Write a description of each sample.

Play Water Game

Play the game, "Water Play." (Your teacher will provide a copy.)

Agriculture and Agribusiness

Field Trip to Natural Water Body

Take a field trip to a nearby body of water. Use a water test kit to test samples from the water source for nitrates, phosphates, and pH. Use the results of the tests to predict problems that might occur. Discuss measures that can be taken to correct the problems.

Different Types of Irrigation

List the different types of irrigation used locally. What are the advantages of each type of irrigation? What are the disadvantages of each type of irrigation? Are there any types of irrigation that are not used locally? Why are these types of irrigation not used?

Health Occupations

Culture Coliform Bacteria from Wastewater

Contact someone at the local wastewater treatment plant. Get from them samples of treated and untreated wastewater. Prepare an appropriate medium to culture coliform bacteria. Have the class grow coliform cultures from the treated and untreated wastewater samples. Discuss the implications to public health of coliform contamination of drinking water.

Effects of Ions Dissolved in Water on Your Health

Discuss the effect of ions often found in drinking water—such as sodium, calcium, magnesium, and fluoride—on health. Find out what ions are present and their concentration. Are any of these beneficial? Do any of the ions in the water present a health threat?

Home Economics

Sodium in Food and Water Supplies

Plan a menu for one day. Using information on product labels, determine the amount of sodium in one serving of each item on the menu. Estimate the water consumed during the day. Find out the concentration of sodium in the drinking water. Calculate the total sodium consumed with this menu. Use a nutrition guide to determine if this amount of sodium is healthy. Discuss the health effects of sodium in the diet.

Testing Swimming Pools and Hot Tubs

Find out how algae and bacteria are controlled in swimming pools and hot tubs. What problems can come from not controlling the algae and bacteria in swimming pools and hot tubs? What problems can occur if the concentrations of the chemicals are too high?

Industrial Technology

Heat-Capacity Values of Different Media

Demonstrate the differences of heat capacities of water, sand, and soil by placing equal weights of each in a small beaker. Measure the initial temperature. Place the three samples under a spotlight for 5 minutes. Turn off the spotlight and measure the final temperature of each sample. The higher the temperature of the material, the lower the heat capacity.

Water has a high heat capacity. Why does this property of water make it valuable as a coolant?

Scale Buildup in Boilers, Etc.

Hard water contains calcium and magnesium ions. What bad effects do these ions have on boilers and water heaters? What can be done to avoid these problems?

HOW DO PLANTS
CONTROL WATER LOSS?

Introduction

The two primary ways that water is returned to the atmosphere during the water cycle are evaporation and transpiration. Plants are involved in and greatly affected by both these processes.

Since a plant is fixed at one location throughout its life, it depends on the soil in which it grows to obtain water. Without sufficient water a plant will grow poorly, its leaves will wilt, and it will die.

Plants lose water in two ways.

1. Plants give up water vapor from their leaves by transpiration.

2. Water evaporates directly from the soil.

These two processes together are called **evapotranspiration**. Evapotranspiration is greatest on hot, sunny days, especially when there is little moisture in the air. If a plant is to grow, it must replace the water it loses by evapotranspiration. Such water is replaced through rainfall or irrigation.

Large farms often employ a crop production technician to determine if fields need to be irrigated. To do this, the technician must find out the following:

1. how much water evaporates out of the soil (done using a device called a tensiometer);

2. how much water a given crop loses through transpiration (done experimentally, as you will do in this lab);

3. how much rain to expect (estimated from rainfall data collected by the local weather bureau) (see Table L4-1).

By knowing how much evapotranspiration and rainfall to expect, the crop production technician can determine how much irrigation is needed.

Purpose

Measure the amount of water a plant loses by transpiration.

Lab Objective

When you've done this lab, you will be able to —

- Relate loss of water by transpiration to conditions in a plant's environment.

Lab Skill

You will use this skill to complete this lab —

- Measure movement of water through a pipette.

Materials and Equipment Needed

a fresh plant cutting	knife
tap water	two-hole rubber stopper
rubber tubing	pinch clamp
1.0-ml graduated pipette	250- or 500-ml beaker
metal elbow	pint mist bottle
electric hair dryer	timer (for minutes)
rubber bulb	lab stand with clamp
plastic "T"	plastic plunger (syringe w/o tip)
narrow-mouthed quart bottle (or 1000-ml flask)	

LAB PROCEDURE

Pre-Lab Discussion

Plants are not entirely at the mercy of their surroundings. Many plants can control their rate of transpiration. The underside of any plant leaf contains thousands of microscopic pores. CO_2 can enter the leaf through these pores and then be converted to food inside the leaf during photosynthesis. But when leaf pores are open, much needed water may be lost from the leaf.

When it is under water stress, a plant can close off some or all of its leaf pores. However, if the plant must keep its leaf pores closed, it can't get CO_2 and isn't making food.

The solution to this problem is to keep a plant well-watered. Water it loses through transpiration is replaced immediately by water from the soil.

In this lab, you will measure transpiration in a plant cutting held in an apparatus like that in Figure L4-1. During the tests you carry out on the cutting, its leaves will lose water. The apparatus measures this water loss.

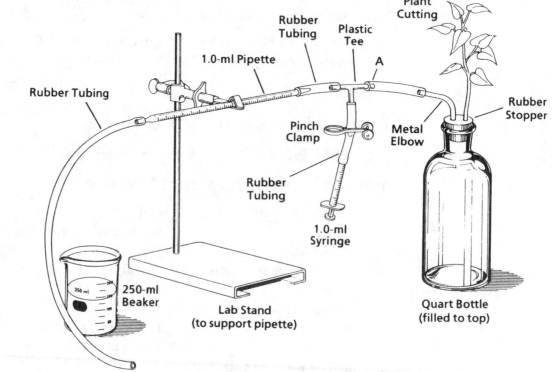

**Figure L4-1
Setup for
measuring
transpiration**

You will be able to measure water loss by your plant cutting under at least three conditions: normal room air, very dry air, and very humid air.

What piece of equipment in the list above would you use to lower the humidity of the air around your plant?

What piece of equipment would you use to increase the humidity?

How will you measure how much water the plant is losing?

Safety Precautions

- Be careful in cutting woody plant stems underwater. Your blade may slice a finger.

- Keep away from water while running the electric hair dryer. Unplug the hair dryer and keep it away from water or mist when not in use.

Method

Setting Up a Transpirometer

1. Fill the 250-ml beaker half-full with tap water.

2. Fill the quart bottle (or 1000-ml flask) completely full with tap water.

3. Assemble the apparatus as shown in Figure L4-1 so the plunger is in the full-out position and the pinch clamp closed.

4. Push down gently on the rubber stopper to make sure the quart bottle is completely filled with water.

5. Obtain a freshly cut stem with numerous healthy leaves:
   - hold the bottom of the stem under water;
   - cut off the bottom one inch.

6. Quickly insert the stem of the cutting through the second hole in the rubber stopper, making sure the fit is watertight.

7. Now disconnect the apparatus at point A in the diagram.

8. Place a rubber squeeze bulb on the free arm of the "T."

9. Place the free end of the long rubber hose into the water in the 250-ml beaker.

10. Slowly squeeze the rubber bulb. (Bubbling will occur in the beaker.)

11. Then slowly release your grip on the bulb and watch the column of water travel up the pipette. (Do not let the water enter the "T.")

12. Undo the pinch clamp and push in the plunger to force the water back to somewhere between 0.0 and 0.5 ml on the pipette scale.

13. Now reconnect the apparatus at point A.

14. Reclamp the pinch clamp tightly.

15. Pull the plunger back to the full-out position.

16. Push down again on the rubber stopper:
    - this should move water in the pipette slightly toward the beaker (if it does not, see your teacher);
    - when you release the stopper, water should start to move very slowly toward the plant.

---

17. Complete Line A of Data Table 1.

    Note: The water level should constantly move toward the plant during tests. If it doesn't, get help from your teacher. You can reset the water level in the pipette whenever necessary by unclamping the pinch clamp and pushing in the plunger.

Measuring Transpiration in Very Humid Air

18. After about 10 minutes, complete Line B of the data table.

    Do each of the steps below without delay in between steps.

19. Use the mist bottle to spray the upper and lower sides of all the leaves on your cutting just until they are dripping wet (leave the plant wet for about 10 minutes).

20. While you wait to complete Step 19, do Calculation F.

21. After the plant has been wet for about 10 minutes, complete Line C of the data table.

Measuring Transpiration in Very Dry Air

22. While your lab partner does Calculation G:

    Keep away from water when you use the dryer.
    - Turn on the hair dryer to the hottest setting.
    - Holding the dryer about 12" from the cutting, dry both the upper and lower sides of all leaves.
    - Move the dryer back and forth to dry the plant evenly. Continue to blow air on the plant for about 10 minutes.
    - Don't expose any one spot of the foliage to heat for more than a second or two.
    - If the leaves begin to show signs of wilting, switch the dryer to the low heat setting or hold the dryer farther from the cutting.

23. As soon as you turn off the dryer, complete Line D of the data table.

24. Allow normal transpiration to take place for 10 minutes. During this 10 minutes, do Calculation H.

25. After ten minutes have passed, complete Line E of the data table.

Observations and Data Collection

Copy the data table below into your ABC notebook and record your results there. DO NOT WRITE IN THE TEXTBOOK.

Data Table

Water Level in Pipette (ml)	Distance Water Level Moved (ml)	Time of Day	Time Elapsed (minutes)	Condition of Plant's Leaves
A.	Do not write in this space.		Do not write in this space.	
B.				
C.				
D.				
E.				

Calculations

Finish the following calculations in your ABC notebook. DO NOT WRITE IN THE TEXTBOOK.

F. Calculate the initial transpiration rate (in ml/hr) in room air in the following way:

$$\frac{\text{Distance Water Moved (Line B)} \times 60 \ \frac{\text{min}}{\text{hr}}}{\text{Time Elapsed (Line B)}}$$

Record the result on the line below:

Initial transpiration rate in room air = _____ml/hr

G. Calculate the transpiration rate (in ml/hr) in humid air using data from Line C.

Record your result on the line below:

Transpiration rate in humid air = _____ml/hr

H. Calculate the transpiration rate (in ml/hr) in dry air, using data from Line D.

Record your result on the line below:

Transpiration rate in dry air = _____ ml/hr

I. Calculate the final transpiration rate (in ml/hr) in room air, using data from Line E.

Record your result on the line below:

Final transpiration rate in room air = _____ ml/hr

Cleanup Instructions

- Remove the cutting and throw it in the trash.
- Take apart the transpiration setup.
- Empty used water down the drain.
- Wash glassware and tubing, and store.
- Return the mist bottle and hair dryer to the lab counter.

WRAP-UP

Conclusions

Answer the following questions in your ABC notebook.

1. Compare the initial transpiration rate (Calculation F) with the transpiration rate in humid air (Calculation G).

   - What does the difference indicate about the opening and closing of leaf pores?
   - What factor caused the plant to react in this way?

2. Compare the initial transpiration rate with the transpiration rate in dry air (Calculation H).

   - What does the difference indicate about the opening or closing of leaf pores?
   - What factor caused the plant to react in this way?

3. Compare the initial transpiration rate with the final transpiration rate (Calculation I). Relate the difference to any change in the condition of the plant's leaves from the beginning to the end of the experiment.

4. Did you record wilting of leaves at any time during the experiment? Explain why this happened in terms of humidity or temperature.

5. How would your normal rate of transpiration have changed if you had used a cutting with twice as many leaves? With half as many leaves?

Challenge Questions and Extensions

Answer the following questions in your ABC notebook.

6. Why would a greenhouse attendant be asked by his/her supervisor to mist plants when the heat is turned on in the greenhouse?

7. Assume the daily transpiration rate of a one-hectare field of tomatoes is 15,000 liters per hectare. A number of tensiometer readings taken in this field during the previous June showed a daily evaporation rate equal to 5000 liters. The average June rainfall is that shown in Table L4-1. How much irrigation water (in liters) must the tomato crop receive in June to keep it from wilting? (Assume it rains the average amount in June.)

Table L4-1 Monthly Rainfall Data

Month	Rain (cm)*	Month	Rain (cm)*
Jan	11	Jul	5
Feb	11	Aug	4
Mar	12	Sep	4
Apr	10	Oct	10
May	8	Nov	12
Jun	5*	Dec	14

* 1 cm of rainfall = 100,000 liters of rain per hectare.

8. According to the rainfall data in Table L4-1, in what months could tomatoes survive without additional water from irrigation?

HOW DOES ACID RAIN AFFECT WATER QUALITY?

PREVIEW

Introduction

You are a fisheries technician in an area of your state where rain frequently has a pH less than 4. The pH of most lakes and streams in your district has fallen sharply over the last ten years. Fish and aquatic plants are especially sensitive to the acid conditions and have been dying in large numbers.

Your State Director of Fisheries has asked you to find a way of restoring some of the lakes that are completely dead. You will need to have an understanding of:

1. how acid rain changes the pH of a lake;

2. how bedrock below the lake may affect its pH;

3. how acids are neutralized by certain alkaline substances.

When sulfuric acid (H_2SO_4) falls from the atmosphere as precipitation, it may enter soil, lakes, streams, or other reservoirs of water. The acid dissolves in the soil or lake water and lowers its pH. These reservoirs, of course, are habitat for many plants and animals.

In some parts of the U. S., the Earth's crust contains rock minerals that help to maintain pH after acid rain has entered a lake. Such rock "buffers" the lake (a buffer is a substance that resists a change in pH). The buffering capacity of bedrock varies greatly from one region to another.

Purpose

In this lab, you will examine how a body of water may be naturally buffered against the effects of acid rain and how an acid-polluted lake can be neutralized.

Lab Objectives

When you've finished this lab, you will be able to—

- Compare the ability of different types of rock to neutralize acidic solutions.
- Test the ability of lime to raise the pH of an acidic solution.

Lab Skills

You will use these skills to complete this lab—

- Measure the pH of a solution using pH paper or a pH meter.
- Crush rock using a mortar and pestle.

Materials and Equipment Needed

scoop or scraping tool	2-inch × 4-inch label cards
tweezers	large funnel
newspapers	160-ml mortar and pestle
500 mg calcium hydroxide (98% purity)	acid rain (25 ml × one plus number of rock samples)
lab apron	100-ml graduated cylinder
pH indicator paper	safety goggles
glass rod	wax pencil
rubber gloves	rock samples
petri dishes (one for each rock sample)	triple-beam balance (121 g @ 0.01 g)

LAB PROCEDURE

Pre-Lab Discussion

If you have not watched your teacher's demonstration of how acid rain is made, do so now. As you watch the demonstration, answer these questions:

1. How can you tell that a gas is actually entering the beaker?

2. How could you tell if the solution in the beaker was actually becoming acidic?

Safety Precautions

- Wear goggles and protective clothing while observing the sulfur dioxide generator and while handling acids.
- Keep acids in a well-ventilated space and do not breathe close to them.

Put on your safety goggles and lab apron now. Leave them on during the entire lab.

Method

1. Get rock samples from your teacher.

2. Use a mortar and pestle to crush rock samples to as fine a consistency as possible.

3. Using a wax pencil, label each petri dish bottom with the rock type it will contain.

4. Scrape each type of crushed rock into a different petri dish bottom and spread the rock evenly.

5. Label each petri dish lid with a label card showing the rock type.

    - List these rock types in column 1 of the data table.

    - Label an extra petri dish "Control."

6. Use a graduated cylinder to measure 25 ml of acid rain from the stock bottle.

7. Pour the 50 ml of acid rain into one of the petri dishes with a rock sample.

8. Stir the sample gently for 10 seconds with the glass rod.

9. Cover the sample with the petri dish lid.

10. Repeat Steps 6 through 9 for each petri dish, including the "Control."

11. Let each sample set for 5 minutes before testing its pH.

12. While the petri dishes are setting, cut twice as many 1/2-inch strips from the rolled pH paper as you have petri dishes. (Don't touch the pH paper with your fingers.)

13. Test the pH in each sample and the "Control."

    - Use the tweezers to place the strip of pH paper completely into the acid rain until it is saturated.

- Remove the strip of pH paper from the acid rain and place it on the label.

- Compare the pH paper to the pH color chart.

- Record the pH of each sample and the "Control" in column 2 of the data table.

14. Set aside each sample with a pH the same as the "Control."

15. To each of these samples and to the "Control":

- Add 0.1 g of $Ca(OH)_2$.

- Stir with a glass rod until the $Ca(OH)_2$ dissolves.

16. Test the pH of each sample again following the directions in Step 13. Record the pH of each sample in column 3 of the data table.

Observations and Data Collection

Copy the data table below into your ABC notebook and record your results there. **DO NOT WRITE IN YOUR TEXTBOOK.**

Data Table

Rock Type	Initial pH	pH After Treatment With $Ca(OH)_2$	Buffering Capacity of Rock
Control			

Cleanup Instructions

- Follow your teacher's instructions for saving any leftover acid rain.

- Use newspapers, a large funnel, and beaker to filter the crushed rock from solutions, rinse the rock with tap water, and dispose of in a solid-waste container.

- Pour all remaining solutions into isolation jars for later disposal. Label the isolation jars.

- Clean all glassware with soapy water, rinse with water, and allow to air dry.

WRAP-UP

Conclusions

1. In column 4 of the data table, rank the rock samples by their ability to neutralize the acid (1 = best, 2 = next best, and so on).

2. Use the data in column 4 of the data table to determine the following:

   - Which rocks seemed the most resistant to neutralization?

   - Could this be explained by how finely you were able to crush the rock?

   - What process in nature might speed up or slow down the buffering of acids by bedrock?

   - What other characteristics of the rock might affect neutralization?

Challenge Questions, and Extensions

Answer the following questions in your ABC notebook.

3. In some areas of the country, soil and water are naturally buffered by bedrock that contains alkaline minerals. These regions are not expected to have acid rain problems even if their rainfall is acid.

   Get a U.S. map that shows major bedrock types throughout the country from your teacher. Based on your lab results, predict which areas of the country might have the most severe acid problems in the environment? Which might have the least?

4.   The calcium hydroxide you used in this lab occurs in limestone. It is called lime by gardeners and groundskeepers who use it to neutralize naturally acid soils. Write the complete chemical reaction that shows how lime neutralizes one of the acids present in acid rain.

5.   Use recent newspaper or magazine articles to find out which states are hardest hit by acid rain problems.

   •   Do these reports agree with what you predicted after examining your results in this lab?

   •   What additional factors related to acid rain might you need to account for when making such predictions?

6.   A fisheries manager wants to reclaim a small fishing pond with a very low pH as a result of a single year of heavy acid rain. He can buy truckloads of lime to add to the pond.

   What information does he need in order to know how much lime to buy?

HOW EFFICIENT IS A WATER FILTER?

Introduction

People and businesses in your community have been complaining for a long time that their drinking water is often cloudy and has a bad odor. A recent analysis of the water revealed the presence of a number of organic substances, as well as suspended material.

The manager of your community's water treatment plant must decide whether to add another step to the treatment process the plant currently uses.

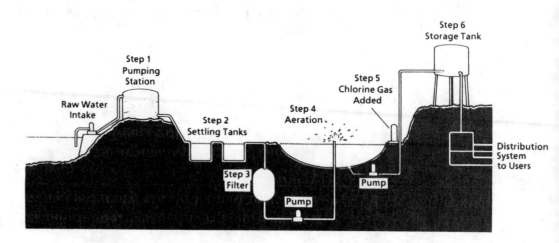

Figure L6-1 Water treatment plant

A final step could be added by installing a filter made of sand or activated charcoal. The plant manager knows that sand is cheaper than charcoal, but will sand get the job done? He has ordered a sand filter and a charcoal filter for a preliminary test. This test will be done in the plant's lab using water samples drawn from the plant's output.

As the plant's lab technician, you will carry out the tests on the water. You will give your results to the plant manager. Using your report and his cost analysis he will decide which type and size filter to install.

Purpose

In this lab, you will compare the ability of two different types of filter to remove a contaminant from a water supply.

Lab Objectives

When you've finished this lab, you will be able to—

- Show how the presence of a solute in water can sometimes be detected.

- Show that a water filter has a limited filtering capacity.

Lab Skills

You will use these skills to complete this lab—

- Use a graduated cylinder to measure liquid volume.

- Use a pinch clamp to adjust the flow rate of a liquid.

Materials and Equipment Needed

500-ml beaker	rubber hose
125-ml beaker	eyedropper
granular activated charcoal	3 10-ml graduated cylinders
pinch clamp	sand
cotton	18-ml test tubes
ring stand	small plastic funnel
long glass rod (18")	lab apron
test tube clamps	test tube rack (6 hole)
safety goggles	hot plate
deionized H_2O	Benedict's solution, 10 ml
10-ml glass or plastic column (8 mm inside diameter)	
0.1% glucose solution, 50 ml	

LAB PROCEDURE

Pre-Lab Discussion

Materials dissolved in water often are not visible. You must test for the presence of such substances using a test reagent that reacts with the substance. Ideally this reaction will give a product with a specific color. Such a reagent (Benedict's solution) will be used in this lab to detect the presence of a contaminant in a water sample. If the water sample is contaminated, it will turn from a blue color, first to green, and then to orange.

CAUTION!

Put on your safety goggles and lab apron now. Leave them on for the entire lab.

Method

1. Set up the apparatus in Figure L6-2 and a boiling water bath (a 500-ml beaker half full of water on a hot plate).

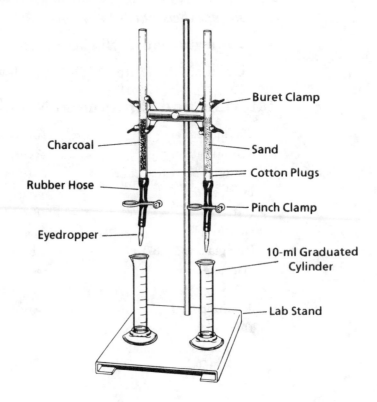

Buret Clamp

Charcoal

Sand

Cotton Plugs

Rubber Hose

Pinch Clamp

Eyedropper

10-ml Graduated Cylinder

Lab Stand

Figure L6-2
Lab setup

2. Use the plastic funnel to fill both the sand column and the charcoal column completely with deionized water. Drain the columns into graduated cylinders.

3. Add 10 ml of each filtrate to test tubes labeled S (for sand) and C (for charcoal). (Set these tubes aside.)

4. Pour approximately 50 ml of contaminated water from the stock bottle into a 125-ml beaker.

5. Perform the following steps:

- Measure out 10 ml of contaminated water from your beaker into one graduated cylinder and pour it into one of the columns.

- Repeat previous step for the second column.

- Wait until the flow rate from a column is 1 drop/sec or slower. (You can adjust the flow rate with the pinch clamp.) Then begin collecting the filtrate in a clear 10-ml graduated cylinder.

6. After both columns have drained completely, add the two filtrates to test tubes labeled 1S (for sand) and 1C (for charcoal). Set these aside.

7. Repeat Steps 5-6, but collect the filtrate in tubes labeled 2S (for sand) and 2C (for charcoal).

8. Label a test tube "K." Pour 10 ml of contaminated water into tube K.

9. Add 5 drops of Benedict's solution to each tube and place all of them into the boiling water bath as shown by your teacher.

10. After tubes have incubated for 5 minutes remove them from the water bath. Complete the data table.

Observations and Data Collection

Copy the data table below into your ABC notebook and record your results there. DO NOT WRITE IN THIS TEXTBOOK.

DATA TABLE

	Color of tube before heating	Color of tube after heating 5 min
Tube S		
Tube C		
Tube 1S		
Tube 1C		
Tube K		
Tube 2S		
Tube 2C		

Cleanup Instructions

- Unclamp the columns from the ring stand and clean them as shown in Figure L6-3.

- Collect the used charcoal and sand in a paper bag or other suitable package and deposit in a trash container.

- Pour both filtered and contaminated water down the laboratory sink drain.

- Wash the graduated cylinders, test tubes, and columns with soapy water and rinse with clean water. Allow these items to air dry. Then return them to their proper storage places.

- Return any unused charcoal or sand to its proper storage place.

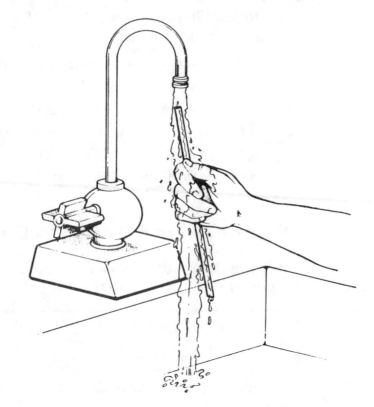

**Figure L6-3
Cleaning
the
columns**

WRAP-UP

Conclusions

Answer the following questions in your ABC notebook.

1. Using the results in the data table answer these questions:

   - How were you able to tell that the water sample was actually contaminated before filtering?

Applied Biology/Chemistry

- Should there have been any color change in tubes S and C after heating? If you did see a color change in these tubes, what could have accounted for it?

- Do color differences between the 1S tubes and the 1C tubes tell you whether sand or charcoal makes the better filter? Explain.

- Do color differences between the 1C and 2C tubes indicate that a filter has a limited capacity? Explain. (If there was no color difference, do a third run with tubes labeled 3C.)

2. All filters eventually become saturated with molecules of contaminant. At this point the filter has reached its adsorption capacity. The adsorption capacity of a filter is the total amount of a contaminant that one gram of the filter can remove.

    The concentration of contaminant in your water sample was 1 gram/liter. (By the way, the contaminant was dissolved sugar!) Can you estimate the adsorption capacity of granulated charcoal? Start by determining how many grams of contaminant you poured into the column on each run.

3. Based on your results and conclusions, recommend the best filter type to use to the plant manager.

4. How do you think the effectiveness of a filter changes with the rate at which contaminated water flows through it? Did you do anything in this lab that kept the water from flowing so rapidly that it avoided filtering?

Challenge Questions and Extensions

Answer the following questions in your ABC notebook.

5. Explain how you could increase the adsorption capacity of a filter by changing the size of its granules.

6. You are trying to remove a pollutant from your kitchen tap water. The concentration of the pollutant in the water supply is 100 mg/liter. You are using a 1000-g activated charcoal filter on your kitchen faucet with an adsorption capacity of 100 mg/gram. The filter is good enough to remove all of the pollutant even when the faucet is fully open. Your use of water for drinking and cooking is about 10 liters per day. How often must you change filters to be sure none of the pollutant gets into the water you use?

7. Design an experiment that would test the ability of charcoal and sand to remove suspended particles from a water sample. If time

permits, carry out the experiment. How do the results of this experiment affect your recommendation to the plant manager in (2) above?

8. An emergency medical technician (EMT) has been called to a home where a young child just drank an unknown organic solvent. The EMT gives the child a glass of water with powdered charcoal suspended in it. Why would the EMT do this?

SOIL AS A NATURAL RESOURCE

THINK ABOUT IT

- Why do you think a hamburger appears at the beginning of a section on soil?

- What are the ingredients of a good hamburger? How do these ingredients depend on soil?

SUBUNIT OBJECTIVES

After you complete this subunit, you will be able to —

1. Describe how soil is formed from organic and inorganic materials.

2. Evaluate how soil layers and soil composition affect soil's ability to support life.

3. Describe how plants are affected by soils lacking in nitrogen, phosphorus or potassium.

4. Explain how minerals are cycled through the environment.

5. Identify three major soil-related problems and their causes.

6. Propose ways to respond to problems related to soil.

7. Identify jobs that are related to soil.

LEARNING PATH

To complete this subunit, you will —

1. Read the text through "What Is the Composition of Soil?"

2. Take part in class discussions and activities.

3. Do Laboratory 7, "Why Is Soil Texture Important?"

4. Read the text through "What Nutrients Do Plants Get from the Soil?"

5. Take part in class discussions and activities.

6. Do Laboratory 8, "How Are Soils Conditioned to Grow Plants?"

7. Read remainder of text.

8. View and discuss the video problem, "Dust Bowl."

SOIL AS A NATURAL RESOURCE

Soil is the loose part of the Earth's crust that gives physical support and chemical nutrition to plant life. Soil is created through both geological and biological processes. Inorganic soil particles are created geologically when rocks disintegrate in various processes known as weathering. Organic soil particles are created biologically when dead organisms decay. Figure 5-1 shows the formation of soil.

If we dig down through the soil, the features of the soil will vary. Its color and texture may change. Often, several layers may be seen. These layers are called soil horizons. The three main layers are the topsoil, subsoil, and bedrock (Figure 5-1).

- Topsoil supports plant life. The topsoil contains 13 of the 16 chemical elements essential for plant growth.
- The subsoil is similar in structure to the topsoil, but it's not rich in nutrients.
- Bedrock is not as loose and porous as the other two layers.

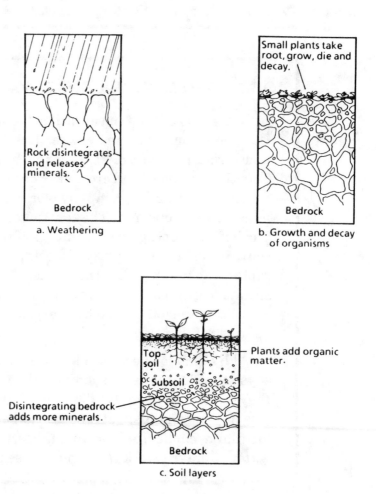

a. Weathering

b. Growth and decay of organisms

c. Soil layers

**Figure 5-1
Soil formation**

Minerals are inorganic chemical substances that occur naturally in the bedrock from which soil is formed. They may eventually become part of the upper layers of soil. Sometimes minerals deposited in bedrock are mined and used to make products. Minerals include these materials found in the Earth:

- Metals

- Ores of metals

- Nonmetal solids

Nonmetal solids include the more solid common components of Earth such as rock, clay, gravel, and sand as well as elements such as phosphorus that plants need as nutrients.

Removing Earth During Strip Mining

Edison Power obtains lignite through strip mining. In strip mining, a strip of earth called overburden is removed above the bed of coal and placed on one side of the coal. The uncovered coal is then removed. A second strip is cut, and the overburden from that strip is placed in the cut from the first strip. Mining continues in this way. The overburden from each new cut is used to fill in the cut that was made before.

The topsoil of the area where Edison Power mines lignite is much better for plant growth than the layers below it. The overburden contains topsoil, subsoil and bedrock. Small amounts of coal and some heavy metals also get into the overburden during digging. Some of these materials are harmful to plants.

By law, Edison Power is required to leave the land in as good or better condition as it was before mining and power plant operations began. The process of returning the land to its former condition or to an improved condition is called reclamation.

ACTIVITY 5-1

With others in the class, make a list of the information about the land that you think Edison Power needs in order to prepare for reclamation.

What Is the Composition of Soil?

Soils are made of inorganic (mostly mineral) particles, organic matter, air, and water. By volume, about 45% to 49% of the soil is mineral, 1% to 5% is organic matter, and about 50% is water and air.

Inorganic Components of Soil

Clay, silt, and sand are the three major types of mineral particles found in the soil. They range in size from small (clay) to medium (silt) to large (sand). The particle size of these components affects the soil's ability to hold water and nutrients.

- Fine sand particles are about 0.10 to 0.25 mm in diameter. Soils with too much sand dry out rapidly. These soils are usually infertile because they don't provide nutrients to the growing plants.

- Silt particles range from 0.002 to 0.05 mm in size. Soils with too much silt tend to separate and erode under heavy rainfall.

- Clay particles are less than 0.002 mm in diameter. Clay holds soil nutrients, determines the acidity of the soil, and holds moisture better than sand or silt. Soils with too much clay tend to form hard clods that inhibit the growth of plant roots.

ACTIVITY 5-2

- Divide into small groups. Your teacher will give each group samples of three types of soil particles as well as mixtures of the three.

- Examine the soil particles. Observe their characteristics.

- Examine the soil mixtures. As a group decide which soil particle is present in the greatest amount and in the least amount.

- Rate the soil mixtures on their ability to hold moisture, nutrients, and forming clods.

- Compare your findings with those of other members of the class.

A good mixture of soil should contain sand, clay and silt. If soil particles are relatively large, the soil is classified as sandy soil. If the particles are very fine, the soil is classified as clay soil. Soil with about an equal mix of sand and clay is called loam. Loam is a good type of soil for plant life.

Ash Containment at Edison Power

One problem at Edison Power is what to do with the ash left over after lignite is burned. Some of the ash will be sold to a company that will use it to make concrete. The remaining ash will be transported to a special type of landfill.

The lignite ash is acidic. It would contaminate the ground water if it were dumped into an ordinary landfill. The landfill first will need to be lined with a three-foot-thick "liner" of soil that will hold in the ash. When the landfill is full, it will need to be covered up with the same material used for the liner. Finally, the landfill will be covered with topsoil and planted.

ACTIVITY 5-3

In your ABC notebook, write a paragraph about which of the three soil types, sand, clay, or silt, would be best for containing the ash from the Edison Power Company. Why?

Organic Components of Soil

Organic matter is material derived from plants and animals. It includes dead and decaying plants and animals, plant parts and animal wastes. Decaying organic matter is called humus. The amount of organic matter, or humus, in the soil to some extent depends on climate. The soils in warmer climates usually have less humus. That's because decay occurs at a faster rate in warmer temperatures. The organic materials are broken down and any inorganic components are released very quickly.

Soils with high content of organic matter have many benefits. These soils will—

- hold water well,

- allow water to move through the soil,

- support microorganisms that promote decay of plants and animals, and

- have the right pH to encourage the release of mineral nutrients.

One way of ensuring that soils have a high content of organic matter is to condition soil through composting. Composting is the mixing of organic matter and other materials with the soil to encourage their decay into nutrients.

Soils Unlimited is a company that develops soils for use in horticulture, landscaping and organic farming. The company collects and transports recycled waste material for large-scale composting. The recycled material is dumped by the ton over several acres of land where the soil is to be developed. The company's production manager, Nathan T., makes sure that the recycled material is spread out in the right amounts. Each type of soil—landscaping, rose-growing, etc.—has its own formula. "The soil is the stomach of the plant," says Nathan, "and our job is to feed it."

The "feed," or recycled material, includes manures (from livestock, horses, poultry, and bats), sawdust from a door factory, and brewery wastes from beer companies. These materials contain organic matter that breaks down into plant nutrients.

ACTIVITY 5-4

- Working with other members of your class, plan a small-scale, backyard composting operation.
  1. What kind of materials would you use?
  2. Where would you obtain them?
  3. How would you move them from wherever they are produced to the compost site?
  4. How much would it cost you to get set up?
- Make a budget, a schedule, and a list of tasks that have to be done.

What Nutrients Do Plants Get from Soil?

Soil is an essential natural resource mainly because it supports most of the plant life on earth. Soil is the basis for almost all agricultural activities (Figure 5-2).

Figure 5-2
Agricultural
uses of soil

Plants need sixteen chemical elements to grow. Among the most important are nitrogen, phosphorus, potassium, carbon, oxygen, sulfur, calcium, iron, and chlorine. Before plants can use an element, it must be placed in a solution that the plant can absorb. For example, nitrogen is one of the most critical elements for plant growth. Plants absorb nitrogen mainly as nitrates (NO_3^-). Nitrates, however, are subject to movement in the soil and to loss when they dissolve in water and are washed away. This process of nutrient loss is called leaching. By contrast, phosphorous and potassium are often "fixed" in the soil in complex compounds that are not readily available to plants.

To grow properly and to produce maximum yields, plants must get a balance of nutrients.

Table 5-1 shows how each of the three primary nutrients affects plants. For each nutrient, you can quickly recognize the following information: form absorbed by the plant, types of growth promoted, symptoms of nutrient deficiency, and symptoms of over-fertilizing.

Table 5-1: Primary Nutrients

Nutrient (element)	Nitrogen	Phosphorous	Potassium
Form absorbed by the plant	Nitrates (NO_3^-)	Phosphates (PO_4^{3-})	Potassium (K_2O)
Types of growth promoted	Dark green foliage and vegetative growth	Sprouting of seeds, faster maturing, and healthy seeds, development of strong root systems	Formation and transportation of food within the plant, healthy, disease-resistant plants, root growth
Symptoms of nutrient deficiency	Sickly yellowish-green color, stunted or slow growth, dried-up leaves	Purplish discoloration of leaves and stems, slow maturing, and low yields	Mottled, spotted, or streaked leaves, which may appear scorched and dried out
Symptoms of over-fertilizing	Too rapid foliage growth, softness of tissue, general plant weakness Burning, brown spots on leaves. Will kill entire plant if extremely excessive.	No symptoms. The plant will take up only the amount it needs.	Increased water content of plants reduces resistance to frost injury, delaying maturity

ACTIVITY 5-5

- Make an illustrated chart showing the symptoms of nutrient deficiency in plants.

- Draw one plant that appears to have a nitrogen deficiency, another that has phosphorus deficiency, another that has potassium deficiency.

- Label each plant to indicate the nutrient problem.

Applied Biology/Chemistry

If plants show any of the symptoms listed in Table 5-1, consider having a soil analysis for a possible deficiency or excess of the associated nutrient. Also, be aware of other possible causes of the symptom, such as deficiency or excess of secondary nutrients, insect attack, and plant diseases. However, if soil analysis reveals a deficiency or excess of one or more essential elements, changing to fertilizers with different proportions of the primary or secondary nutrients may solve the problem.

ACTIVITY 5-6

On every fertilizer package, whether it is fertilizer for outdoor or indoor plants, you will find a label such as the one that appears below.

- Read the label and find out: What is the percentage of nitrogen, phosphorus, and potassium in the fertilizer that is represented by this label?

- Do you have any fertilizer at home? Check the label for percentages of primary nutrients.

- Report back to the class.

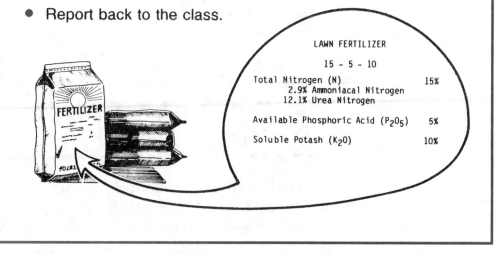

LAWN FERTILIZER

15 - 5 - 10

Total Nitrogen (N)	15%
2.9% Ammoniacal Nitrogen	
12.1% Urea Nitrogen	
Available Phosphoric Acid (P_2O_5)	5%
Soluble Potash (K_2O)	10%

Biogeochemical Cycles

Directly or indirectly, animals get their mineral nutrients from plants. When plants die and decay, the minerals that are in them are recycled in the soil. Minerals in animals are recycled the same way.

The processes through which minerals are recycled through the environment are known as biogeochemical cycles. Figure 5-3 is an example of the phosphorus cycle. Plants incorporate phosphorus into organic compounds that are passed along to the animals that eat the plants. As plants and animals die, microorganisms return the phosphorus to the soil through the decay process. Through erosion and mining, phosphorus may make its way to the sea. Much of the phosphorus that finds its way to the sea is bound up in the deep sediments of the ocean floor where it is removed from the biogeochemical cycle for millions of years.

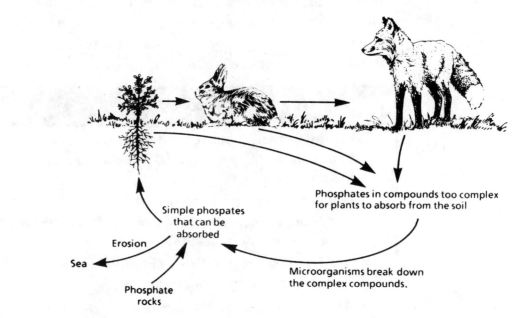

**Figure 5-3
Phosphorus
cycle**

Phosphates in compounds too complex
for plants to absorb from the soil

Simple phospates
that can be
absorbed

Erosion

Sea

Phosphate
rocks

Microorganisms break down
the complex compounds.

What Problems and Issues Are Related to Soil?

Is soil a limited or unlimited resource? There's lots of it, for sure. But soil is not fixed permanently in one location. It moves around. At a shoreline, the sand shifts easily day after day. As it becomes rich with life like kelp and weeds, it becomes harder to move. The fact that soil moves around leads to one of the major problems related to soil: erosion. Two other significant problems are the loss of soil nutrients and changes in the pH of the soil.

Erosion

When soil is lost from one place and moves to another, we call it erosion. Figure 5-4 shows erosion caused by both water and wind. When rivers overflow their banks during a flood, the rapidly flowing waters rip up the soil and take it downstream. As the current

---

becomes calm, the soil settles out. Even during normal rainfall, erosion can wash literally millions of tons of soil annually into lakes and streams. Sometimes this can be very harmful to both plant and animal life. Erosion can also be caused by wind, which was the case in the "Dust Bowl."

a. Wind

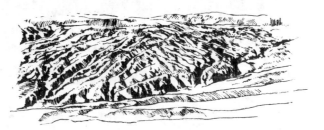

**Figure 5-4
Erosion**

b. Water

ACTIVITY 5-7

You now know something about the strip mining and power plant operations at Edison Power.

- Explain how the activities listed below cause erosion. Consider how the weather might contribute to erosion.
    - Strip mining
    - Construction of roads for transporting coal
    - Power plant construction
- Write in your ABC notebook ways to control these erosion problems?

Nutrient Depletion

The problem of nutrient depletion can be caused in several ways. Too much rain can cause leaching of vital elements. Leaching is the process in which nutrients dissolved in water are washed from the soil.

Nutrient depletion can also occur when the same crop is grown year after year. Identical crops grown over a period of time can use up certain elements for which they have a particular need. Eventually, the soil is depleted of those nutrients and it no longer can support plants.

In recent years, tropical rain forests in South America have been cleared to allow conversion of the land for farming. Soil in a tropical forest is not rich with nutrients because trees readily absorb them. Farming further depletes the soil's nutrients. The soil is not suited for replenishment of nutrients. After a few years, the tropical farmland crops fail and the farms are abandoned. The result is loss of two major resources—the rain forest and the soil.

ACTIVITY 5-8

- Find an area around your home or school where the soil has been disturbed by some outside influence.

- Watch what happens to the soil for several days.

- In your ABC notebook, keep a record in the form of a log or diary in which each observation is dated. Each entry should include the following information: soil conditions (dry, crumbly, wet, coarse, eroding, etc.), weather conditions, signs of life in the soil, human activities that have affected the soil.

- Organize the information into a chart when you have completed your observations.

Changes in Soil pH

Earlier in this unit, you learned about acid rain. One of the effects of acid rain is that it can alter the pH of the soil and make it acidic. Changes in pH may drastically affect the availability of plant nutrients. Most plants, therefore, don't thrive in soil that is very acid or very alkaline. Figure 5-5 shows the pH range suitable for growth of various fruits and vegetables. The pH of soil can also be affected by the types and amounts of fertilizers that a farmer or gardener may use and chemicals used to control pests or weeds.

pH RANGE

	3	4	5	6	7	8	9	10
BLUEBERRIES		4.0||||||||||||| 5.5						
CRANBERRIES		4.2|||||| 5.0						
POTATOES			4.8||||||||||||||||| 6.5					
STRAWBERRIES			5.0|||||||||||||| 6.5					
PEPPERS			5.5|||||||||||||| 7.0				
CAULIFLOWER			5.5||||||||||||||||||| 7.5				
TOMATOES			5.5|||||||||||||||||| 7.5				
ONIONS				6.0||||||| 7.0				
PEAS				6.0|||||||||||| 7.5				
ASPARAGUS				6.0|||||||||||||||||||| 8.0			

Figure 5-5 pH range of fruits and vegetables

ACTIVITY 5-9

You are a farmer in the northeastern United States. Assume your land is properly watered and fertilized with a pH of 5.5.

● Answer these questions in your ABC notebook:

1. What crops in Figure 5-5 would you plant?

2. What crops might be best to plant if acid rain further lowers the pH of your soil?

3. What is the pH of the soil around your home? How could you find out this information?

What Might Be Done About the Problems?

Erosion

Erosion problems may be dealt with in several ways.

● Dams can prevent rivers from ravaging land around them.

● Terraces can be built in fields to inhibit the flow of water so it won't wash away the topsoil (Figure 5-6).

● Contour plowing is another way to control erosion on hilly and sloping fields. In this method, the fields are plowed so the furrows run perpendicular to the direction of the slope

---

(Figure 5-6). This practice slows runoff as it flows down the slope.

**Figure 5-6
Preventing
erosion**

a. Terracing   b. Contour Plowing and Windbreaks

- Leaving the stubble from harvested crops in a field can help prevent both water and wind erosion.
- Planting cover crops that serve primarily to hold the soil in place can help.
- Leaving natural vegetation on the land can also help prevent erosion.
- Fields can be protected from wind erosion by rows of trees that serve as windbreaks (Figure 5-6).

Nutrient Depletion

The problem of nutrient depletion is being addressed in several ways.

- One way is by alternating the kind of crops grown on a field from one year to the next. This is called crop rotation. Because different crops have different nutrient requirements, crop rotation ensures that no one nutrient in the soil will be overused.
- Depleted nutrients in soil can be partially replaced by spreading fertilizers. These may be manufactured chemicals or simply animal manures, which are especially rich in phosphorus.
- Clearing land of all vegetation can deplete soil nutrients. This practice is giving way to removing only the most economically valuable plants. Selective removal does little damage to soils, and it helps conserve nutrients, both in the soil and in the tissues of the uncut plants.

Changes in pH

We also need to be very careful about putting fertilizers and pesticides in our soils since these chemicals can alter soil pH. There is a healthy market now for food products that are grown by organic methods. With organic farming methods, growers don't use manufactured chemicals to fertilize the soil or to control pests and weeds.

Soils that are too acidic to grow most crops can be reconditioned. Lime (calcium hydroxide) can be added to soils to neutralize their acidity.

ACTIVITY 5-10

Here are three occupations that depend upon or affect soil.

- Analyze each occupation listed below in relation to soil. How is soil involved?
  - Potter (craftsperson who makes pottery)
  - Landscape architect
  - Building contractor

Looking Back

Soil is the loose part of the Earth's crust that gives physical support and chemical nutrition to life. Soil is made from **inorganic materials** (substances that contain few or no carbon atoms in their molecules) as well as **organic** materials (substances in which carbon atoms form the center or backbone of the molecules). Soils with high organic matter content hold water well, allow water to move through the soil, support microorganisms, and have the right pH.

The three main layers of soil are **topsoil**, **subsoil**, and **bedrock**. **Clay**, **silt**, and **sand** are the three major types of mineral particles of soil.

Plants need sixteen chemical elements to grow. Nitrogen, phosphorus, potassium, carbon, and oxygen are among the most important. Most of these elements are located in the soil. Plants take these nutrients up through their roots and make them into food for the plant to grow. Animals eat the plants, and as animals die and decay, the nutrients are released back into the soil.

The three major soil-related problems include erosion, nutrient depletion, and changes in soil pH. Solutions include changes in farming practices and conditioning of soils. Many jobs exist to deal with these problems.

Vocabulary

The words and phrases below are important to understanding and applying the principles and concepts in this subunit. If you don't know some of them, find them in the text and review what they mean. They're listed in the order in which they appear in the subunit.

inorganic	nitrogen
weathering	nitrates
organic	phosphorus
soil horizon	phosphates
topsoil	potassium
subsoil	biogeochemical cycle
bedrock	erosion
clay	leaching
silt	contour plowing
sand	crop rotation
loam	fertilizer
humus	pesticide
composting	organic farming

Further Discussion

- Call or visit a plant nursery in your community. Report to the class about the different soils required to grow indoor and outdoor plants in the nursery. How many different kinds of soil mixtures can you identify?

- What is the difference between soil and dirt?

Activities by Occupational Area

General

Local Soil Profile

Get a county soil survey map from the local Soil Conservation District. Identify from the survey the types of soil where you live and around the school. Compare the soil type, classification, slope, and recommended use listed on the survey to the actual properties and uses.

Soil Percolation Test

Contact an inspector from the county health department. Invite the inspector to speak about "soil perc tests" performed to determine the suitability of soil at a building site for a septic tank and the number and type of lateral lines needed.

Demonstrate the difference in permeability of sand, silt, and clay by packing a funnel with each type of soil. Pour 100 ml of water through each funnel and measure the time needed for the 100 ml of water to run through each sample.

Agriculture and Agribusiness

Soil Sample

Collect a soil sample from the school campus. Describe the sample as moist or dry, clay or sand. Describe the physical characteristics of the site of the soil sample such as flat, gentle slope, or steep slope. Note any evidence of erosion at the site. Discuss any changes in land use at the site that could correct any problems noticed.

Wind and Water Erosion of Soil

Set up pans of soil samples at different grades such as 5°, 15°, and 30°. Pour 250 ml of water at the top of the slope of each sample. Compare the results for the three samples.

Set up three samples of fine soil in a pan. Leave one soil sample without a windbreak. For the other two samples, prepare a windbreak using 4-inch lengths of ¼-inch dowel. Position the three samples in front of an electric fan. Have the windbreak of one sample between the soil and the fan and for the other sample, have the windbreak on the end of the pan away from the fan. Turn on the fan and let it blow on the three samples for five minutes. Compare the three samples.

Health Occupations

Diseases in the Soil

Identify diseases that are caused by bacteria and fungi in the soil. List the symptoms and treatment of each disease. How can these diseases be prevented?

Measurement of Iron in Blood

Use a hemoglobin scale to measure the iron in a sample of artificial blood.

Home Economics

Houseplants Grown in Deficient Soils

Grow some houseplants in soil samples that are deficient in different nutrients. Notice the effects of nitrogen deficiency, phosphorus deficiency, and potassium deficiency. Add a fertilizer with the needed nutrient to some of the plants with each of the different deficiencies. Notice the changes in the plants given fertilizer as compared to the plants left deficient in that nutrient.

Facts and Fallacies of Organically Grown Food

Research the facts and fallacies of organically grown food. What health claims are made about organic foods? What benefits can you expect from organic food?

Industrial Technology

Sand, Gravel, and Clay Use in Building Materials

Discuss how soil is used in building materials for general construction. Components of the soil are important resources in different types of construction.

Ores and Mining

In strip mining, more than the ore is removed from the ground. Discuss the responsibility of mining companies for restoring mining sites. How do mining companies restore mining sites?

WHY IS SOIL TEXTURE IMPORTANT?

Introduction

You are a soil technician working for your county's soil conservation service. Part of your job is to control erosion on local farms and recreational lands. Your supervisor, the County Soil Conservation Agent, has given you the task of making a map showing the quality of all soils in the county. You are mainly concerned with soil factors that determine a soil's ability to hold water and resist erosion. One of these factors is known as soil texture.

Soil texture refers to the ratio of different-sized mineral particles in the soil. Based on size, mineral particles are referred to as clay, silt, or sand. (The size ranges for clay, silt, and sand particles were given earlier in the subunit.)

Clay particles are small and electrically charged and attract and hold water molecules. Thus, it is important that soils used to grow crops have some clay content. Clay also helps make soil resistant to erosion.

Like clay, silt particles are also small enough to hold water in the soil. But exposed silt washes away easily, taking nutrients with it.

A soil with too much sand content allows water to pass through very quickly and does not hold enough moisture to support most plants. Exposed sandy soils may be subject to erosion if the angle of the land, the slope, is too severe.

Purpose

In this lab, you will determine the texture of a soil sample you collect.

Lab Objective

When you've done this lab, you will be able to —

- Analyze the texture of a soil to determine the percents of clay, silt, and sand.

Lab Skills

You will use these skills to complete this lab —

- Separate a mixture of particles.
- Determine the ratio of different size mineral particles in a soil sample.

Materials and Equipment Needed

1 half-pint bottle with cap
tap water
soil sample
newspapers
large funnel
spatula or spoon
cardboard

LAB PROCEDURE

Pre-Lab Discussion

A soil technician needs to know about the behavior of soil particles. The rate at which soil particles settle out of a liquid depends on their size. Larger particles settle faster than smaller particles. How could you use this fact to determine the texture of a given soil sample?

Safety Precautions

- Be sure the cap is screwed on tightly before you shake the bottle.
- Dry your hands before handling the bottle.

Method

1. If you have not already done so, collect enough soil to fill a gallon-size plastic bag.

    - Collect the sample from the top 2-4 inches of soil.

    - Label the sample with your name, date, and collection site.

    - Also on the label, describe the following features of the collection site:

      Slope   Amount and type of vegetation
      Moisture level How the land is being used

2. Crumble your sample into small pieces, breaking up any large clods.

3. Spoon enough loose soil from your sample to fill your bottle half full.

4. Fill the bottle almost to the top with tap water.

5. Cap the bottle and shake it.

    - When your sample is well mixed, place the bottle upright on the lab table.

    - Describe the appearance of your soil sample on Line A of Data Table 1.

    Allow the mixture in your half-pint bottle to settle until distinct layers of soil have formed beneath the water.

Note: Some soils may take several hours to settle, particularly if they include large amounts of clay. If your sample hasn't formed distinct layers after 20 minutes, label it and keep it until the next lab period. Repeat Steps 2-6 using a different soil from elsewhere in the lab.

7. Hold a piece of cardboard next to the half-pint bottle.

    - Draw horizontal lines on the cardboard to mark the soil layers.

    - Next to each layer write the type of soil particle found in that layer: clay, sand, silt.

    - Perform the calculations in the next section to fill in Lines B, C, D, and E of Data Table 1.

Observations and Data Collection

Copy the data table below into your ABC notebook and record your results there. **DO NOT WRITE IN THIS TEXTBOOK.**

DATA TABLE 1

A. How did your sample look after you mixed it with water?

_____

_____

B. % sand   _____%

C. % silt   _____%

D. % clay   _____%

E. Soil type for your soil sample   _____

Do the following calculations to obtain Lines B, C, D, and E of this table.

Calculations

Do the following calculations in your ABC notebook.

1. Using the formula below, calculate the % of each type of soil (sand, silt, clay) in the soil column of your half-pint bottle.

$$\% \text{ sand} = \frac{\text{Depth of sand layer in mm}}{\text{Total depth of soil in mm}} \times 100\%$$

2. Using the %s in Data Table 1 and the soil triangle in Figure L7-1, determine the soil type for your sample as follows:
   - Find each % on the proper axis of the soil triangle (Figure L7-1).
   - Follow the dotted lines from each % to the point where they intersect inside the soil triangle.
   - Identify your soil type by identifying the region of the soil triangle that contains the point of intersection. (For example, for a soil with 50% sand, 30% silt, and 20% clay, the 3 dotted lines intersect inside the region marked "loam.")
   - If the dotted lines intersect on the boundary between two regions, assign the soil type of the graph region of greater area.

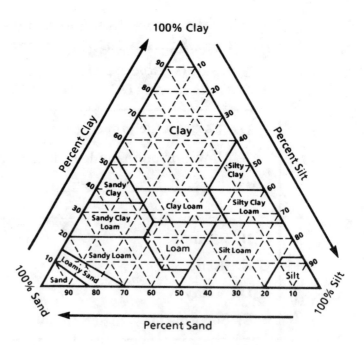

**Figure L 7-1
Soil triangle**

Cleanup Instructions

1. Save your soil sample for use in another lab.

   • Seal the plastic bag containing the remainder of your soil sample.

   • Add the soil type to the label you made for your soil sample in Step 1 of the lab procedure. (Be careful to attach the label to the bag so that it won't be lost.)

   • Store the sample as directed by your teacher.

2. **Do not empty your half-pint bottle directly into the sink. (Pouring soil directly into the sink will stop up the drain.)** Empty the bottle only as follows:

   • Form a filter cone from several sheets of newspaper.

   • Place the cone in a large funnel.

   • Place the funnel in the sink drain.

   • Empty your half-pint bottle into the cone.

   • Fill the bottle with clean water and again empty it into the cone.

   • Repeat the rinsing process as many times as needed to remove all of the soil particles from the bottle.

   • After the water has drained, throw the newspaper and used soil into the trash.

3.   Clean and store the funnel and the half-pint bottle.

   - Wash with soapy water.

   - Rinse with clean water.

   - Dry.

   - Store in the proper place.

Conclusions

Answer the following questions in your ABC notebook.

1.   As each lab group in the class tells the soil type it has identified, place a mark inside the corresponding region of the soil triangle.

   - Do the marks fall mainly in one area of the triangle?

   - From these results, which type of soil particle is most common in your region?

2.   Do some samples on the triangle fall outside the main cluster of samples?

   - If so, what factors at the collection site are associated with those types of soil?

   - Do you think these factors are the cause or the result of the soil texture at that site? Explain.

3.   Which two soil samples in the class could be mixed together to obtain the best loam?

Challenge Questions and Extensions

Answer the following questions in your ABC notebook.

4.   Predict the answers to the following questions:

   - Which of the soil samples in the class will hold the most water?

   - Which of the samples will be most subject to erosion?

5.   Plan simple lab experiments to test the predictions you made in Question 1.

6.   If time permits, carry out the experiments you planned in Question 2, as directed by your teacher.

HOW ARE SOILS CONDITIONED TO GROW PLANTS?

Introduction

Mr. Smith is a salesman for a landscaping company. Last year he filled the local high school's order for ten dozen red cedar bushes. The groundskeeper at the school just called him and said that few of the cedars have grown despite lots of sunshine and rain last year.

"Those were our best stock of red cedar," Mr. Smith tells the groundskeeper. "With all the rain we've had this spring, they should have grown at least twelve to eighteen inches. I think we may have a problem with the soil. I'll come over this afternoon and run some tests. We may have to add some nutrients to the soil to condition it."

The nutrients Mr. Smith referred to are the primary nutrients required by all plants—nitrogen, phosphorous and potassium. They can be added to soils in various forms of fertilizers that you will learn about in this lab.

Purpose

Test the condition of a soil by running pH, nitrogen, phosphorus, and potash tests with a soil test kit.

Lab Objective

When you've finished this lab, you will be able to—

- Condition a soil so that it will support the growth of a given type of plant.

Lab Skills

You will use these skills to complete this lab—

- Apply a test reagent to a sample using a dropper.
- Relate the color change of a soil solution to its pH or nutrient content.

Materials and Equipment Needed

soil sample   safety goggles
soil test kit (Sudbury or similar)   lab apron
paper towels   dropper
spatula or small spoon   newspaper

Pre-Lab Discussion

The tests that Mr. Smith referred to will determine what amounts of the three primary plant nutrients (N, P, and K) are present in the soil. The tests are done using a soil test kit obtained at any garden shop or nursery.

As you will see in this lab, a soil test kit is easy to use. The results it gives tell a lot about how well a certain species of plant will grow in the soil being tested. For example, if the test indicates that the soil is deficient in one or more nutrients, the test kit provides information that will help you calculate the amount of fertilizer to add.

The fertilizers in Table L8-1 are commonly used to make up nutrient deficiencies. Each fertilizer has a rating of three numbers. These numbers show the % of N, P, and K it contains. Study the table and determine which fertilizer you would use to supply each of the three main nutrients, N, P, and K.

Table L8-1: Nutrient Content of Various Fertilizers

Fertilizer	% Nitrogen	% Phosphorus	% Potash
Fish scrap	8	13	4
Blood meal	15	1	0.5
A 5-10-5 fertilizer*	5	10	5

* For a bag of commercial fertilizer, the % of N, P, and K will be listed in that order on the front of the bag.

The type of soil test kit you will use also tests the soil's pH. Think about why pH is important in properly conditioning the soil.

Safety Precautions

- **DO NOT** point stoppered test tubes toward anyone. The solutions in the test tubes may be under pressure and blow the stoppers off. If vigorous bubbling occurs in the test solution, loosen the stopper slightly to release pressure.

Method

1. Find the instruction booklet inside your soil test kit.

2. In the booklet, find the procedures for the following four tests:

   - pH test
   - Nitrogen test
   - Phosphorus test
   - Potash (potassium) test

3. For each of the four tests, follow the kit's instructions exactly. (This lab can use the Sudbury Soil Test Kit, or any other kit that is available.)

Observations and Data Collection

Copy the data table below into your ABC notebook and record your results there. DO NOT WRITE IN THIS TEXTBOOK.

DATA TABLE 1: pH and Nutrient Tests

A. pH test = _____

	Nutrient Concentration	% Nutrient Requirement
B. Phosphorus test	_____ppm	_____%
C. Nitrogen test	_____ppm	_____%
D. Potash test	_____ppm	_____%

Cleanup Instructions

1. **Remember not to pour any water containing soil directly down the drain.**

   - Dump any soil left in test tubes into a paper towel.
   - Rinse test tubes until all soil particles are removed, always pouring your rinse water through a newspaper funnel into the drain.
   - Throw paper towels and soil into the trash.

---

2. Wash all glassware in soapy water—then rinse, air dry, and store in proper places.

3. Return the reagents, test tubes, and tin stirring rod to the soil test kit.

WRAP-UP

Conclusions

Answer the following questions in your ABC notebook.

1. Look at the results of the soil nutrient tests you ran in this lab.
   - In which of the primary nutrients was your soil sample most deficient?
   - If you grew a plant in this soil (without adding fertilizer), what problems or symptoms would the plant be likely to show?

2. Look at the results of your pH test.
   - Did your soil sample have a pH that was acid, alkaline, or neutral?
   - Red cedar grows well in a soil whose pH is 5.5–7.0. If your soil had no nutrient deficiencies, could it be used to grow red cedar? (If not, you will determine how to condition it in Question 7.)

3. Examine the results of all nutrient tests run in your class.
   - Determine the primary nutrient that was most often deficient in the soil samples.

4. Construct a graph like the one below.
   - Enter each group's data from this lab and Lab 7 on the graph.
   - Using the data on the graph, try to relate a deficiency of this nutrient to the soil type determined in Lab 7.

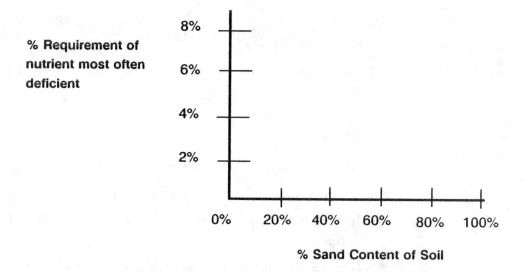

% Requirement of nutrient most often deficient

% Sand Content of Soil

5. Construct a second graph and try to relate a deficiency of this nutrient to some feature at the site where the soil was collected. (For example, if your class measured slope where it took its soil samples, your graph might look like the one below.)

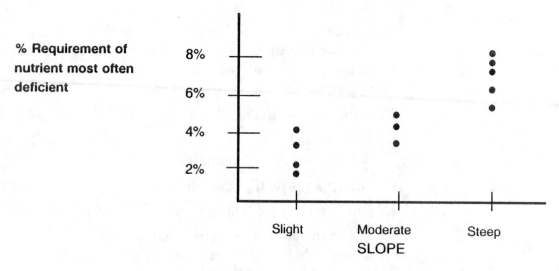

% Requirement of nutrient most often deficient

SLOPE

6. In the sample graph above, what process accounts for the nutrient becoming more and more deficient in the soil as the slope increases?

7. Using Tables L8-1 and L8-2, determine how much of each fertilizer (fish scrap, blood meal, or 5-10-5) is needed to meet the % requirement of nitrogen for your soil sample.

Applied Biology/Chemistry

Table L8-2: Pounds of Fertilizer Needed for 1,000 sq ft

% Requirement of N, P, or K	% N, P, or K in your fertilizer:						
	2%	3%	4%	5%	6%	8%	15%
2%	50	33	25	20	16	13	7
3%	75	50	38	30	25	19	10
4%	100	66	50	40	33	25	13
6%	150	100	75	60	50	38	20
8%	200	132	100	80	67	50	27

Challenge Questions and Extensions

Answer the following questions in your ABC notebook.

8. What rating on a 50-lb bag of fertilizer would exactly make up for all your soil sample's nutrient deficiencies? (Hint: Use Table L8-2 again.)

9. Your soil sample was used to fill a 5' x 20' shrubbery box of red cedars.

   - What substances in Table 8-3 below could you add to obtain the correct pH?

   - How many pounds of each substance would you add to each shrubbery box?

10. Design an experiment to determine how a deficiency of one of the primary nutrients would affect plant growth.

Table L8-3: How to Change the pH of a Soil.

A. To lower the pH by one unit, add: (All additions are given in pounds per 1000 sq ft.)

For soils:	Less than 30% Sand	30-55% Sand	Greater than 55% Sand
Compost	150-200	100-135	67-90
Animal manure	50	34	23
Sewage sludge	50	34	23

B. To raise the pH by one unit, add:*

For soils:	Less than 30% Sand	30-55% Sand	Greater than 55% Sand
Ground limestone	78	68	45
Spray lime	58	50	33
Burnt lime	44	38	25

* Lime should not be applied in amounts greater than 40 lb/1000 sq ft at one application.

LIVING RESOURCES: PLANTS AND ANIMALS

THINK ABOUT IT

- The giant panda, a native of China, is an endangered species. Its primary source of food, bamboo, has been drastically reduced by agriculture, development, and harvest for products. What happens when an animal's food source is destroyed?

- What factors do you think might cause an animal or plant to become extinct?

- Whose job is it to protect endangered species? What occupations might be involved?

- What part do nature preserves and zoos play in saving plant and animals species?

SUBUNIT OBJECTIVES

After you complete this subunit, you will be able to —

1. Identify the ways that people use plants and animals as natural resources.

2. Identify uses of plants and animals as being consumptive or nonconsumptive.

3. Explain how plants and animals are related through the food web and through the carbon dioxide-oxygen cycle.

4. Explain the major problems that affect plants and animals as natural resources.

5. Propose ways to respond to problems related to plants and animals.

6. Discuss jobs that are directly involved with plants and animals.

LEARNING PATH

To complete this subunit, you will —

1. Read the text through "What Problems and Issues Are Related to Plants and Animals?"

2. Take part in class discussions and activities.

3. View and discuss the video problem, "Animal House."

4. Do Laboratory 9, "Does the Effect of Acid Rain on Crops Depend on Soil Factors?"

5. Read the remainder of the text.

6. Take part in class discussions and activities.

7. Do Laboratory 10, "How Should We Manage Our Game Species?"

8. View and discuss the video "Natural Resources: A Summary."

9. Complete the Unit Wrap-Up Activity.

Living Resources: Plants and Animals

Edison Power's Effects on Plants and Animals

The proposed Edison Power Plant will be built in a prairie environment. The plants and animals living on the prairie are important resources for the people in the area. As Edison Power mines the coal, the plant life of the prairie will die as it is covered by mounds of overburden. Animals that depended on these plants for food or shelter will either move out of the area or die.

ACTIVITY 6-1

You are a citizen of Richmond.

- Think about how plants and animals may be used as resources for your community.

- Write a letter in your ABC notebook to the editor of the Richmond newspaper as if you were one of the following persons: a dairy farmer, a hunter, a birdwatcher, or a nature photographer.

- Express your concerns in the letter about what you think will happen to the plants and animals in your area as a result of the company's mining and plant operations.

Plants and animals are resources that are widely distributed on Earth. The places where they live, their habitats, include other resources discussed in this unit: air, water, and soil. Plants are an important part of an animal's habitat.

Different kinds of living organisms are called species. Members of a species have a set of common characteristics. When they mate and reproduce, these characteristics are passed to their offspring. Some species have characteristics that make them valuable as resources. For example, bamboo, a tall grass grown in Eastern Asia, is valued for its strong but flexible stems. Bamboo can be used to make scaffolds, fishing poles, and cane chairs. It is the primary food source of the giant panda.

Why Are Plants and Animals Important?

You are probably not aware of all of the uses of plants and animals. Just a few of the more obvious uses are shown in Figure 6-1. We can classify all the uses of living resources into one of two types: consumptive and nonconsumptive. When trees are cut to give wood for furniture, we are consuming (using up) trees, so our use is consumptive. When maple trees are tapped for syrup, the trees do not die. So our use of them is nonconsumptive. What's important is that nonconsumptive uses of resources do not deplete the resources.

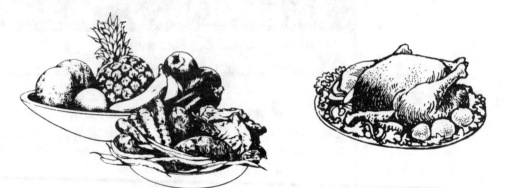

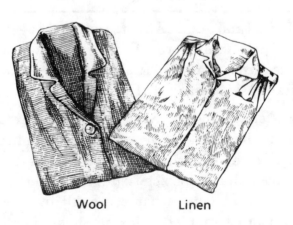

Wool   Linen

**Figure 6-1
Uses of plants
and animals**

Ways That People Use Plants and Animals

Plants and Animals Provide Food. Some plants and animals supply us with food. Fruits, nuts, and vegetables provide sources of sugar, starch, oils, roughage, and vitamins. Fish, poultry, livestock, and wild game are important sources of protein and minerals.

Plants and Animals Provide Raw Materials for Products. A list of plants and animals from which we harvest raw materials would fill a book larger than the one you're reading. Just a few of the major raw materials taken from plants and animals are given in Table 6-1.

Table 6-1: Plants and Animals as Sources of Raw Materials

Raw Material	Comes from	Use
Hemp	Plant Fibers	Burlap, Rope
Flax	Plant Fibers	Linen
Tannin	Plant Leaves	Dyes
Resin	Pine Sap	Lacquers, Adhesives
Gum	Sap	Food Additives
Rubber	Sap	Tires
Morphine	Fruit	Pain-killer
Digitalis	Dried Leaves	Heart Medicine
Silk	Caterpillars	Clothing
Hide	Animal Skin	Leather Goods
Serum	Animal Blood	Vaccines, Antivenoms

Plants and Animals Contribute to Human Health. Many drugs used as medicines come from plants. Drugs such as quinine, opium, morphine, digitalis, cocaine, and atropine are plant extracts. Cough medicines, laxatives, antacids, vitamins, and salves also come from plants. Agar, an extract of seaweed, is used in hospital laboratories as a medium for growing bacteria.

Animals are in constant use in medical research. New medical techniques such as organ transplants are often developed first with animals. Testing of drugs, food additives, and other potentially harmful chemicals is done on rodents and monkeys. Insulin for diabetics and antivenoms for snakebites are products derived from animal tissues.

ACTIVITY 6-2

Medicines are very important in our lives. They have extended our lives dramatically, and have greatly improved the quality of life.

- Following the example of Table 6-1, create your own table in your notebook. Put the following medicines in the raw materials column and complete the "comes from" and "uses" columns:
  - quinine
  - aspirin
  - penicillin

Ways That Plants and Animals Are Related

The species of plants, animals, and other organisms that live in a given region form a community. This community of life and the nonliving world with which it interacts form an ecosystem. The processes that go on in ecosystems link plants and animals closely. You saw this in the biogeochemical cycle in Figure 5-3 in Subunit 5. Two important processes that drive ecosystems, photosynthesis and respiration, are carried out inside plant and animal tissues.

Photosynthesis. Plants and algae have special pigments for trapping the energy from sunlight to make their own food. One such pigment, called chlorophyll, is a green molecule found in plant leaves. During photosynthesis, chlorophyll absorbs sunlight. The energy gained from sunlight is stored in carbohydrate molecules being made within the plant's leaves. The carbohydrates are organic molecules made from the carbon dioxide in the air and the water in the plant leaves. As the carbohydrate is formed, the plant releases oxygen to the air.

Photosynthesis can be summarized in a chemical equation.

Photosynthesis:

$$6\,CO_2 \;+\; 6\,H_2O + \text{energy from the sun} \longrightarrow C_6H_{12}O_6 \;+\; 6O_2$$

(carbon   (water)   (carbohydrate)   (oxygen)
dioxide)

Respiration. You might think of respiration as inhaling and exhaling air, but it also has another meaning. Respiration is a process that takes place in every tissue of an organism. It occurs in plants as well as animals. It's how living tissues get energy from their food to grow, reproduce, and carry out other special functions.

In respiration, carbohydrates release their energy when they react with oxygen. It's a little like the combustion of coal. Compare the equation below for respiration to the one for combustion.

Respiration:

$$C_6H_{12}O_6 \;+\; 6O_2 \longrightarrow 6CO_2 \;+\; 6H_2O + \text{energy}$$

(carbohydrate)   (oxygen)   (carbon   (water)
                                               dioxide)

Hydrocarbon Combustion (Butane):

$$2C_4H_{10} \;+\; 13O_2 \longrightarrow 8CO_2 \;+\; 10H_2O + \text{energy}$$

(butane)   (oxygen)   (carbon   (water)
                                              dioxide)

Photosynthesis and respiration equations are also useful in describing two ways plants and animals are related: 1) through the food web (Figure 6-2) and 2) through the carbon dioxide-oxygen cycle (Figure 6-3).

Food Web. Plants and animals are related through the carbohydrates in the chemical equations for photosynthesis and respiration above. Carbohydrates represent the most useful form of energy for organisms. Carbohydrates made by plants are sources of energy for animals. Animals that feed on plants receive this energy. As they become the prey of other animals, called predators, the energy of carbohydrates made in photosynthesis is passed on again. Such food relationships between organisms in an ecosystem are referred to as a food web.

A food web also makes clear that every species has a critical role in an ecosystem. For example, the size of a corn crop in a field will depend on the number of grasshoppers eating the leaves of corn plants (Figure 6-2). The number of grasshoppers will depend on how many mantids and lizards eat grasshoppers. The number of lizards, in turn, is controlled by the number of hawks eating lizards. Remove any of these species and the numbers of any of the others will be affected.

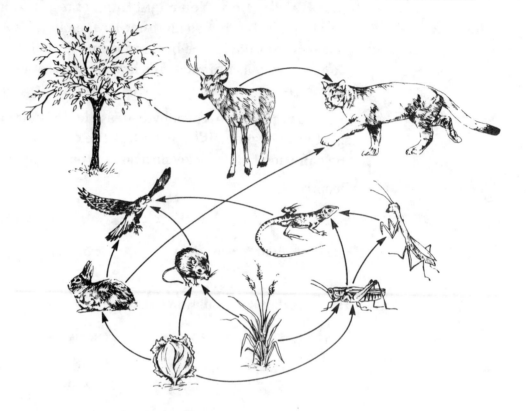

**Figure 6-2
Food web**

Carbon Dioxide-Oxygen Cycle

Look again at the chemical equations for photosynthesis and respiration discussed earlier in this section. What substance is produced in respiration that's used again by plants? What substance is produced in photosynthesis that is needed by both plants and animals? The relationship between the two substances you named can be shown in the carbon dioxide-oxygen cycle (Figure 6-3).

Animals depend on the oxygen produced by plants during photosynthesis. Plants depend on the carbon dioxide produced by animals during respiration. People also depend on the food web and the carbon dioxide-oxygen cycle for their survival.

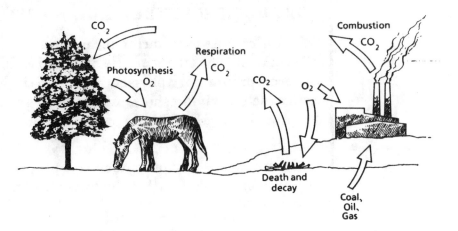

**Figure 6-3
Carbon dioxide-
oxygen cycle**

What Problems and Issues Are Related to Plants and Animals?

Because plants and animals reproduce themselves, they are renewable resources. Does this mean they are unlimited? Or are we in danger of running out of oak trees, wheat, silkworms or sheep?

Let's look at three major concerns related to plants and animals:

1. keeping plants and animals free from natural enemies;

2. making plants and animals more productive;

3. ensuring plants and animals are renewable.

JOB PROFILE: VETERINARIAN TECHNICIAN

John W. is a veterinarian technician. He works with a veterinarian who treats mostly dogs and cats. He sometimes performs preliminary physical exams on the animals. Other duties include giving shots, taking blood and carrying out simple lab procedures. His favorite part of the job is assisting with surgery.

John also has a lot of contact with pet owners. He makes appointments, takes patient histories, and dispenses information about pet care.

John went through a two-year training program at the local community college to prepare to be a veterinarian technician. He says, "The training was essential; I had to have it to get the job. But what I like is that I've continued to learn, every day, on the job."

Pests

A pest is considered an organism that has become so abundant that it is threatening the economic value or even the survival of some species that it uses for nourishment or shelter. Such pests may be harmful directly or indirectly to humans. Some animals have become pests of plant or animal products that we use. Moths eat our clothes, termites eat our houses, and rodents eat our food and garbage.

One example of a pest is the gypsy moth caterpillar. The gypsy moth was originally imported to the United States from Europe, to produce a more economical, disease-resistant breed of silkworms. Some gypsy moths escaped from their breeding cages. With no natural predators to eat them, gypsy moth populations grew rapidly. Gypsy moth caterpillars damage forest trees by consuming leaves from trees.

A pest may also be a parasite, an organism that lives on or inside another organism and may damage it in some way. Typical parasites include viruses, bacteria, fungi, and many worms. The organism in which the parasite lives is the host. Parasites may cause disease in some members of a host species. Over time, the host species as a whole comes to tolerate a parasite. But when parasites are newly introduced to an area, they often present problems.

An example of a parasite is the Dutch elm fungus. This parasite gets into a tree's watery tissues through a cut in the bark. It causes leaves to wilt and fall and can eventually kill the tree. European elm trees were mostly resistant to this tree-killing fungus. But when the fungus arrived in North America with imported European logs in 1930, it attacked healthy American elms, which began to die. Some North American shade tree varieties of elm have almost become extinct. New varieties have been produced commercially to replace them.

Oak Wilt

Oak trees are highly valued as shade trees in our neighborhoods and parks. But some species of oak have become highly susceptible to a parasite called the oak wilt fungus. This fungus produces tiny spores that attach to insects or are carried by the wind. Spores from a diseased tree enter a healthy tree through openings in the bark. The spores then grow inside the tree's watery tissues. This cuts off water from the roots to the branches. Branches eventually fall as the tree dies a slow death.

Plant experts called tree surgeons are often requested by homeowners to prune branches or cut down trees. This involves climbing trunks with special spiked shoes.

ACTIVITY 6-3

- Write a paragraph in your ABC notebook advising tree surgeons how to avoid spreading oak wilt.
- Compare your advice to the advice recommended by others in your class.

Productivity

Growing and harvesting crops on an annual basis is a major role of agriculture. Optimum plant growth often requires fertilizing and irrigating the soil. But fertilizers can degrade other natural resources, and irrigation may help deplete local water supplies.

One way of increasing crop productivity is by combining the desirable traits of two parent varieties. The resulting offspring is called a hybrid. A hybrid corn, for example, has been developed that needs less water than parent varieties. Hybrid livestock have been produced that give more meat or milk. There is, however, a limit to how much any plant or animal species can be improved by breeding.

Another problem with production is that domestic varieties may be competing with wild species. Plants grown as row crops compete not only with one another, but also with weeds, for nutrients and water. Domestic animals share pastures or rangeland with wild animals. They may compete with them for food or become their prey. The destruction of sheep and calves by coyotes in the western U.S. is an extreme case of the latter problem.

Renewability

Plants and animals will be renewable if—

1. their habitats are maintained so they can obtain food and water, shelter, mates, and protection from predators;

2. there are enough healthy organisms in a population capable of reproducing.

A plant or animal population that's unable to renew itself will die out. This may mean it disappears from a local area or that it becomes extinct (disappears entirely from the Earth). Approximately 400 species of U.S. plants and animals are currently in danger of complete extinction. Another 120 are on the threatened list. The endangered groups in Table 6-2 have populations so small that they will die out if no steps are taken to save them. Threatened species are those whose populations are beginning to decline. They will become endangered if nothing is done to help them renew themselves.

Table 6-2: Numbers of Endangered and Threatened Plants and Animals in the United States[2]

Group	Endangered	Threatened
Plant Species	149	40
Animal Species	248	82
Mammals	50	7
Birds	76	10
Reptiles	15	18
Amphibians	5	4
Fishes	47	30
Snails	3	5
Clams	31	1
Crustaceans	8	
Insects	10	7
Arachnids	3	

[2]U.S. Department of Interior, Fish and Wildlife Service, January 1989

There are three major reasons for extinction: degradation of habitat, overharvesting and pollution.

ACTIVITY 6-4

- Ask a fish and game manager in your area what local species are endangered. What human activities are responsible for the decline of the species?
- Identify ways people in the community can help restore the species.
- Report your findings in your notebook to share with your class.

Disturbing or Destroying Habitat. When land is cleared for human enterprise, many wild species vacate. In the winter of 1989, for example, the number of elk in Yellowstone National Park declined dramatically. Part of the prime elk habitat was lost to development in and around the park. Complicating this situation was the damage done to the habitat by the tremendous wildfires in the park during the summer of 1988.

What is habitat disturbance for some species may be an ideal change for others. For example, white-tailed deer thrive where forest vegetation has been replaced with shrubs.

Overharvesting. Domestic species of plants and animals are rarely in danger of being overharvested. Crops can be renewed from commercial seeds; eggs are continuously hatched by breeder hens. But if we remove wild animals faster than their populations can be replaced by reproduction, the species will decline. At some point a species may become so rare that it can be wiped out by a catastrophe such as a flood or freezing weather.

JOB PROFILE: FOREST TECHNICIAN*

Jim Sheridan is a special agent for the U. S. Fish and Wildlife Service. Here is how he explains his job: "I'm a police officer for wildlife. I went to the Federal Law Enforcement Training Center (FLETC), which is a federal police academy. Every year I attend 40 hours of school to brush up on law enforcement and I have to pass a target test with a pistol. I have all the training of any other police officer, except I protect wildlife.

"How do I do that? Well special agents keep a close eye on animals and animal parts coming into and leaving the state. Congress made it against the law to transport protected and illegally obtained wildlife across state lines. That law is called the *Lacey Act*. So, some special agents spot check airport baggage. We also patrol in the field. Since Alaska is so big and the wildlife refuges are so remote, about half of our agents use small planes for patrols. Sometimes agents conduct a *stakeout*. A stakeout is when they stay hidden and wait in a certain area where they think laws are being broken. It's hard catching law-breakers in the act.

---

*Courtesy of Alaska Department of Education, Office of Adult and Vocational Education, Juneau, Alaska 99811

"Sometimes I work *undercover*. Undercover means I act like I'm a hunter just like everyone else. That way I see to it that hunters or hunting guides are obeying the laws. Sometimes my job is dangerous. Special agents have to be prepared for a lot of things—for unexpected trouble, or for wilderness survival.

"Special agents also do lots of paperwork. I work on budgets, order supplies and review permits. True, the job is exciting but also pretty routine, too.

"I've always had a feeling for wildlife. Nearly all the special agents do. Lots of people would like to be special agents like me. I think that's because a lot of people want to do what I do to protect wildlife."

An extreme example of overharvesting is the clear-cutting of tropical forests. Clear-cutting removes all of the vegetation in a patch of forest (Figure 6-4). Huge tracts of forests near the equator are being clear-cut to make way for farming.

Figure 6-4
Clear-cutting
of forests

Clear-cutting in the tropics has also caused many species living in equatorial forests to become extinct. Might some of these species have been of use to us? Perhaps some unknown forest tree or shrub would have provided us with a drug against cancer or arthritis.

Clear-cutting also causes soil erosion and leaching of nutrients from tropical soils. The removal of vast areas of photosynthetic organisms increases the overall amount of carbon dioxide in Earth's atmosphere. It also reduces the amount of oxygen returned to the atmosphere by plant respiration.

Environmental Pollution. Pollution can be a cause of species extinction. Individuals of a species can tolerate only certain levels of any harmful substance. We are just now learning what such levels are for people. We have very little information about toxic levels for plants and animals.

DDT in the Food Web

The bald eagle is an example of how water pollution can affect a food web. These stately birds rely on fish for much of their diet. But the fish they ate contained an insecticide known as DDT. For years DDT had been running off farmland into lakes. It was first taken up by algae then passed along the food web to the eagle's prey and then to the eagle.

Figure 6-5 shows the concentration of DDT in the tissue of species at each level of the food chain. The numbers in parentheses represent how many times more concentrated the DDT is from the point of entry into the food web. The concentration of DDT is expressed as parts per million (ppm).

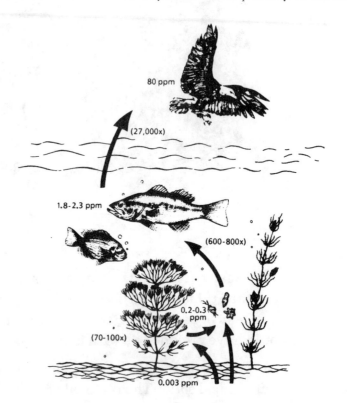

80 ppm

(27,000x)

1.8-2.3 ppm

(600-800x)

0.2-0.3 ppm

(70-100x)

0.003 ppm

**Figure 6-5
Concentration
of DDT through
a food web**

In the case of DDT, trouble was not detected until the top of the food web (the eagle) had been reached. Substances such as DDT, which are stored in body tissue because they cannot be broken down, become more concentrated at each level of a food web. DDT concentrated in the eagle's tissues caused it to make eggs with very weak shells. The eggs would crack before hatching. The eagle's reproductive rate dropped so rapidly that it was declared endangered.

JOB PROFILE: ANIMAL CARETAKER*

Jerry Deppa is the president of the Alaska Raptor Rehabilitation Center in Sitka. He describes his work. "At the center we take care of eagles, hawks, owls and other raptors who are injured. Raptors are birds of prey. Raptors hunt other animals. The center is like a hospital for raptors. Right now we have 14 bald eagles, two young goshawks and one western screech owl in the center.

"I used to be a wildlife biologist. Others who work here all have other jobs. Some of them work as EMTs, doctors or nurses. Still others work in stores, offices or at the pulp mill—all kinds of jobs. Our veterinarian kindly donates his time. It's wonderful how people help save the eagles—and other raptors.

"But on the other hand, it's awful to see what people can do. Some of our eagles have been shot; others have been caught in traps; still others have been shocked on electrical wires or strangled on swallowed fish hooks. Eighty percent of the birds' injuries are human-caused.

"We mostly feed fish to our raptors. Our goal is to get these birds healthy so they can be free again. Sometimes the veterinarian sets a broken wing. Sometimes one of the workers bathes an oiled eagle. Sometimes I take the hawk out to a guywire we designed—for a practice flight. It takes time to get the birds healthy again.

---

*Courtesy of Alaska Department of Education, Office of Adult and Vocational Education, Juneau, Alaska 99811

"One of our best days was last week when we let an eagle go. From all over Sitka people gathered at the beach. When we let the bird go, the eagle waited a moment, almost unbelieving, then flew up to a branch. He looked down at us for several minutes, then took off. He flew out to sea, then turned, and flew right by us twice, as if to say 'thanks' and then was off. That made me happy. Everybody clapped."

Pollutants also affect plants. High concentrations of sulfur dioxide in the air destroy a plant's ability to make food through photosynthesis. Other air pollutants such as ozone, hydrocarbons, and nitrogen oxides also interfere with plant growth.

ACTIVITY 6-5

- Contact Friends of the Earth, the Sierra Club, or another environmental organization to obtain information on another pollutant or additive that interferes with plant or animal growth or survival.

- Make a diagram or flow chart to show how plants and animals are affected by the substance you investigated. Your chart may be similar to the one in Figure 6-5, or you may use more cartoonlike drawings, symbols or boxes to represent the animals and plants.

- Share your chart with members of the class.

- As a class, discuss these questions:
  1. What factors should be considered in weighing the benefits of some pesticides and additives against their harm to plants and animals?

  2. Do some substances seem to concentrate more than others? Which ones?

What Might Be Done About These Problems?

Controlling Pests

One major way of controlling pests has been through the use of pesticides. But few pesticides have worked well. One problem is that many helpful species are killed by the chemicals used against the pest. Another is that not every organism in the pest population exposed to the pesticide dies. Some are naturally resistant. They pass on their resistance to their offspring when they reproduce. As a result, the pest population remains as large as ever, but comes to consist entirely of pesticide-resistant "superbugs" (Figure 6-6). To knock out the superbugs, we try bigger doses of the same pesticide or newer, more powerful pesticides. Some pest organisms will be resistant even to these. The cycle continues.

Figure 6-6
Superbug

Another way of controlling pests has been through quarantine. This involves inspecting living organisms or materials, such as shipments of imported elm logs, to make sure they contain no pests or parasites.

Biological methods of pest and disease control are being developed. These biological methods include producing disease- and pest-resistant varieties of crops so superior to old ones that natural enemies become extinct.

Biological control also includes promoting natural enemies of the pests. Imported pests may have few enemies in their new habitat. Predators must be brought in to control them. One example is the way in which scale insects were controlled in the early 1900s. Scale insects had been ruining orange crops in the U.S. because they had no enemies to control them naturally. The ladybug was imported to eat the scale insects and was successful in controlling them.

Chemical and biological control methods are most often used together in an overall control plan. Such integrated pest management is one way of limiting the amount of harmful pesticides getting into food webs.

Solving Problems Related to Productivity

A number of alternatives to conventional agriculture are being tried. Biologists are hoping to give crops the ability to convert nitrogen in soil directly into nitrates for more rapid growth. Other plant species are being biologically altered to carry out photosynthesis more efficiently. This alteration should increase their yield.

JOB PROFILE: FOREST TECHNICIAN

A forest technician gathers basic forest data, such as determining species and populations of trees, wood units available for harvest, disease and insect damage, tree seedling mortality, and potential forest fire hazards. He/she collects data from instruments, such as rain gauges, thermometers, stream flow recorders, and soil moisture gauges. Additionally, forest technicians train and lead conservation workers and give instructions to visitors concerning forest regulations.

Some forest technicians get on-the-job training, and many are trained through college programs. Forest technicians may work in state and national forests and parks as well as for private industry, such as logging companies.

Some animals are economically valuable for some highly desirable trait. A new technique, known as embryo transplant, is being used to increase the number of offspring from such animals. As soon as a valued cow is pregnant, the embryos are taken from her womb. They are then transplanted into the womb of another cow. The second cow is called a surrogate. As the offspring are being raised by the surrogate, the true mother can be mated again.

Renewing Living Resources

To preserve living resources, the causes of extinction must be controlled.

Habitat. The first step in preventing any species from becoming extinct is to safeguard its habitat or restore habitat that is degraded. For example, strip mining completely destroys habitat. Reclaiming the land, however, encourages most plant and animal life to return. Natural preserves can be set aside in areas of high human activity to safeguard habitat.

Some species whose prime habitat has been all but lost are considered so valuable that they have been preserved through cultivation. An example is a species of periwinkle, a small plant that grew in large numbers in Madagascar's tropical environments. Clear-cutting destroyed its habitat. The chief reason for cultivating this species is that it provides a chemical used in treating a form of leukemia, a type of cancer.

Reclamation of Edison Power's Mined Land

Since Edison Power is concerned about its overall effect on the Richmond area, it plans to replace living resources on the mined prairie. By law, reclaimed land must be as productive as the land before mining takes place. The land, however, need not be replanted for the same use as the original. So the residents of the region will have the opportunity to decide what the new prairie will look like.

ACTIVITY 6-6

- Consider what factors will be important to each of the following occupations as they recommend what to plant on the reclaimed land: dairy farmer, sheep rancher, tomato grower, beekeeper, forester, game warden. . .

- Discuss as a class how the land should be restored to meet the needs of each of these occupations or other occupations you might think of. Can it be restored to meet the needs of more than one group?

Overharvesting. Careful management of crops, forests, and wild animal populations should ensure that they remain renewable for future generations of users. For example, hunting laws that outlaw the taking of female animals just coming into sexual maturity ensure that the species can maintain reproduction to balance population losses.

Pollution. As more biological methods of pest control are developed, the number of polluting chemicals used in agriculture should decline. In the U.S. laws have been passed that require manufacturers of new chemicals to have them tested for adverse health effects. Most cities have strict new codes for the quality of their air and water.

Looking Back

People use plants and animals as natural resources. Plants and animals provide us with almost unlimited uses, including food, raw materials to make products, and medicines. All uses of living resources can be classified as one of two types: **consumptive** or **nonconsumptive**. Consumptive uses of living resources are those uses that harvest and destroy. Nonconsumptive use of natural resources does not deplete the resource.

Plants and animals are related through the **food web**, with every species dependent on each other. There is also a **carbon dioxide-oxygen cycle**. In this cycle, animals depend on oxygen produced by plants, and plants depend on carbon dioxide produced by animals.

Major problems that affect plants and animals include pests, decreasing productivity, and threats to renewability. Solutions to these problems include advanced fertilizers, safe pesticides, renewing endangered species, protecting habitats, avoiding overharvesting, and eliminating pollution.

Vocabulary

The words and phrases below are important to understanding and applying the principles and concepts in this subunit. If you don't know some of them, find them in the text and review what they mean. They're listed in the order in which they appear in the subunit.

habitat	host species
species	parasites
consumptive	hybrid
nonconsumptive	extinction
ecosystem	endangered species
photosynthesis	threatened species
respiration	overharvesting
food web	clear-cutting
predator	pollution
prey	quarantine
carbon dioxide-oxygen cycle	biological control
pests	integrated pest management

Further Discussion

- Plants are often used to make our environment more livable. Animals are used as sources of pleasure or sport. Discuss some ways you and your family benefit from plants or animals for pleasure or sport. Which uses are consumptive or nonconsumptive?

- Make a class list of all of the parks, wildlife sanctuaries, grasslands, state and national forests, or other sites in your area that are set aside for recreation and/or conservation of plant and animal life. How many have you visited? Ask a representative from one of these sites to come to the class and discuss employment opportunities.

- Debate the pros and cons of protecting all species of plants and animals from extinction. For instance, should we care about endangered rattlesnakes? Kangaroo rats? Milkweed?

Activities by Occupational Area

General

Vegetation in Environment

List the naturally occurring vegetation in the local area. Discuss the use of each form of vegetation. Does the use of a natural resource necessarily have to result in its consumption? Classify each entry in your list as consumptive or nonconsumptive.

Food Web

Collect pictures of plants and animals. Organize these pictures into a food web. Discuss the effect on the web of an increase in one of the animals in the food web. Discuss the impact on the food web of spraying pesticide on one of the plants.

Agriculture and Agribusiness

Key Species of Trees

Collect leaves from 20 different trees. Identify each of the trees from its leaf. Mount the leaves and identify the trees. Write a short discussion of the economic and environmental significance of each tree.

Production of Lean Animals

Americans have become more conscious of dietary fat and cholesterol. Discuss how this has affected beef and pork production.

Small-Scale Hydroponics Setup

Set up and run a small-scale hydroponics tank. Determine the advantages and disadvantages of hydroponics on a commercial scale.

Health Occupations

Pharmacist Speaks on Drugs Derived from Plants

Invite a pharmacist to speak to the class about drugs that are derived from plants.

Animal-Borne Diseases Affecting Humans

Discuss animal-borne diseases that affect humans. How can these diseases be prevented?

Home Economics

Vegetarian Diet

Plan menus for a vegetarian diet. Be sure to get the correct variety to provide all of the essential amino acids. Discuss the advantages of this diet.

Animal Rights and Cosmetics and Furs

Research how animal-rights groups are affecting the cosmetic and fur industries. Collect articles and advertisements from both sides: manufacturers of these products and animal rights groups. Form two teams and debate the issue.

Industrial Technology

Properties of Woods Used in Furniture and General Construction

List woods used in furniture construction. List woods that are used in general construction. Compare the characteristics of the woods in the two lists. Compare the characteristics of the trees that produce the different woods.

DOES THE EFFECT OF ACID RAIN ON CROPS DEPEND ON SOIL FACTORS?

Introduction

Phil Sanders works at an Agricultural Experiment Station in a county where the economy depends heavily on cultivating crops. The region has been receiving increasing amounts of acid rainfall. Local farmers have reported lower crop yields each of the last three years. They are worried that low crop yield could be due to acid rain. If so, can they bring yields back to their former levels. They have asked Phil if adding more fertilizer or liming the soil would offset the effects of acidity. If not, they would like the experiment station to develop an acid-resistant variety of crop.

To answer these questions, plant and soil scientists at a nearby university have designed some experiments to be conducted at the experiment station. Research technicians like Phil Sanders will carry out the experimental treatments and keep records of the results. The experiments will be done both in the station's greenhouses and in outdoor test plots. The conclusions of the studies will be published in the station's monthly bulletin and circulated to county extension agents across the state. The agents can then advise farmers about new techniques that will help them improve crop yields.

In this way, farmers are able to deal with the many problems that continually beset even the most modern agricultural systems: pests, lack of rain, need for better equipment, and even environmental pollution.

Purpose

In this lab, you will look at how soil factors affect plants growing in "acid rain."

Lab Objective

When you've finished this lab, you will be able to—

- Determine whether plant nutrition has any effect on a plant's ability to withstand acid rain.

Lab Skills

You will use these skills to complete this lab—

- Grow plants from seeds, or transplant seedlings to a new soil.
- Water and mist plants.
- Test pH of soils and solutions.
- Test nutrient levels of soils.
- Identify symptoms of stress and nutrient deficiency in a plant.
- Weigh plant parts and measure growth.

Materials and Equipment Needed

clear plastic wrap or sheeting	wax pencil or marker
soil test kit	drying oven or desiccator
2 plant misters	scissors
4 plant trays	safety goggles
metric ruler	lab aprons
pH paper	sulphuric acid solution (pH 3.0-3.5)
soil sample	carbonic acid solution (pH 5.7)
(to fill two plant tray)	triple-beam balance
potting soil	watering can
(to fill two plant trays)	soybean seeds or seedlings
nylon (net) bags	

LAB PROCEDURE

Pre-Lab Discussion

In this lab you will carry out an experiment to test the effects of acid rainfall on the growth of an important crop plant, soybeans. Research on crops like soybeans has shown that the acid falling on plant leaves is perhaps more harmful than the acid reaching the soil.

You will test the combined effect of acid rain and soil factors (nutrient levels and pH). The results could be used to answer the question: Can plants receiving extra nutrients overcome the destructive effects of acid?

> **Safety Precautions**
>
> - You will be handling acids in this lab. Be careful not to splash them on your clothes or skin.
>
> - When misting plants with acid, be sure to spray away from people and any materials that could be damaged, such as furniture, books and clothing.

Put on your safety goggles and lab apron now. Wear them whenever you are in the lab.

Method

You will conduct this experiment over several weeks.

1. Obtain four trays. Label each with your name and soil sample label. Mark one of them A, one B, one C, and one D.

2. If you start with seeds follow procedure "a" below. If you start with seedlings follow procedure "b."

   a. Complete the steps below to start seeds:

   - Fill trays A and B with potting soil to within 3/4 " of the top of trays.

   - Fill trays C and D in the same way but with soil sample.

   - Plant as many seeds as possible in each tray at a distance of 4" to 6" apart. Plant seeds to the depth indicated on the seed packet.

   b. Complete the steps below to transplant seedlings:

   - Fill trays A and B 1/4 full with potting soil.

   - Fill trays C and D 1/4 full with soil sample.

   - Dig up seedlings one at a time from the large potting tray. Transplant each to trays A, B, C, and D until trays have as many seedling as possible 6" apart.

   - Add potting soil to trays A and B and soil sample to trays C and D until trays are filled to within 3/4" of top.

   - Cover soil around the plants in each tray with plastic wrap to protect the soil from the rainwater you will spray.

In your ABC notebook, record the date you planted your seeds or transplanted potted seedlings.

3. Place your seedlings or seeds under fluorescent lights or on a brightly lit window sill. Soybeans will do best with 16 hours a day of bright light.

4. Obtain 500 ml of carbonic acid solution and 500 ml of sulphuric acid solution in separate plant mist bottles. Also fill a watering can with carbonic acid solution.

5. Label the carbonic acid bottle "Normal Rain" and the sulphuric acid bottle "Acid Rain."

6. Test the pH of the solutions in both mist bottles and in the watering can. Record the results in your ABC notebook.

7. **When using the plant mister, be very careful to direct the mist only toward the plants. Remember, there is acid in the mist.** When seedlings are 14 days old, mist them three times a week until the end of this unit as follows: (Make sure all trays have their soil protected with plastic wrap.)

   Mist trays A and C with "Acid Rain."

   Mist trays B and D with "Normal Rain."

   Mist all plants with a fine spray until the leaves and stems are dripping wet.

   Water the soil under the plastic wrap as needed with the watering can. Water just enough to keep the soil surface slightly damp throughout the entire course of the experiment. (You can check the dampness by pressing your finger into the soil.)

8. Keep a log for the four trays in your ABC notebook to record the following:
   • Dates of misting and watering
   • Height of plants
   • Number of leaves per plant
   • Length of three largest leaves on a plant
   • Color and general appearance of plants' leaves

9. As plants drop leaves, collect them from each tray and place them in nylon bags labeled A, B, C, and D.

10. When directed by your teacher. Use a soil test kit to test the pH and N, P, and K levels of the soil in all trays.

11. When directed by your teacher, do the following:

- Remove plastic wrap from all trays.

- Cut stems of all plants at soil level with scissors.

- Place cut plants from each tray in bags labeled A, B, C and D.

- In a drying oven or desiccator, completely dry fallen leaves and cut plants, keeping plant matter from the four trays separate and labeled.

- When plant matter has thoroughly dried, weigh fallen leaves and cut plants, keeping them separate and labeled. Record these weights in your ABC notebook.

Cleanup Instructions

- Dump any sickly or dead plants and soil into newspaper or plastic garbage bags, wrap them up, and discard in the trash. Your teacher will tell you what to do with your healthy plants.

- Make a filter cone of newspaper. Rinse all soil particles from the empty trays into the filter. Fold the paper up and discard in the trash.

- Wash the plant trays with soapy water, rinse thoroughly, and store as directed by your teacher.

- Pour unused acid solutions into the sink.

- Wash and store the watering can and plant misters in their proper place.

WRAP-UP

Conclusions

Answer the following questions in your ABC notebook. DO NOT WRITE IN YOUR TEXTBOOK.

1. What general conclusions, if any, can you make about the effects of acid rain on soybean growth.

- Use all of the data you recorded in your ABC notebook:

  – number of plants

  – length of leaves

  – color and general appearance of leaves

  – weight of dried fallen leaves and dried cut plants

- Does this effect depend on soil factors? Which ones? Explain.

Applied Biology/Chemistry

2. Do you think any of the primary nutrients were deficient to the point that they affected plant growth? If so, which ones? Why do you think so?

Challenge Questions and Extensions

(Attempt only if you have completed Question 3 in Lab 7.)

Answer the following questions in your ABC notebook.

3. Compare your results to the results obtained by others in your class. Try to explain differences based on soil pH or nutrient levels. If you can't, what other factors might contribute to the differences?

4. How do the results of your final soil tests on trays C and D compare to the tests you ran on the same soil in Lab 8? How can you explain these differences?

5. Design an experiment to test the combined effect of acid rainfall on soybean leaves with acidity of the plant's soil.

HOW SHOULD WE MANAGE OUR WILDLIFE SPECIES?

Introduction

Cliff Davis is wildlife manager at a state wildlife refuge. His primary job is to ensure favorable conditions for the health and survival of two main wildlife species—deer and grouse.

Cliff and his team periodically survey game habitat to see that it provides adequate food and shelter for the species. They also collect data on wildlife populations that allow them to estimate the number of animals alive at a given time. They gather these data by spotting animals from trails (called censusing), charting nesting areas, and recording litters, as well as kills and other factors responsible for death (mortality factors).

Cliff performs one of his most critical jobs during hunting season when local hunters roam the preserve in search of deer or grouse. His wildlife personnel are on constant vigil to see that hunters do not exceed the bag limit, and that they do not harvest individuals vital to the survival of the population, especially mature females and strong males.

Harvesting, if not managed, is one of several factors that could lead to the elimination of wildlife from the refuge. When losses from a population cannot be balanced by additions, a species declines in numbers. The smaller the population gets, the more easily it can fall to a catastrophe, such as drought, fire, or disease. Another risk in a small population is that its members tend to be closely related to one another. Mating among such animals is termed inbreeding and often results in hereditary disorders and generally weaker animals.

Purpose

To show how proper management of a wildlife species can help maintain the size of the population.

Lab Objective

When you've finished this lab, you will be able to —

- Simulate the effect of reproduction, migration, and mortality factors on the size of an animal population.

Lab Skill

You will use this skill to complete this lab —

- Graph the growth or decline of an imaginary animal population.

Materials and Equipment Needed

checkerboard—covered with clear contact paper

poker chips or checkers (at least 25 each of two colors)

cardboard screen (same size as checkerboard)

pair of dice

graph paper

wax pencil or dry-erase markers

"impact" cards

"wildlife management" cards

paper towels

LAB PROCEDURE

Pre-Lab Discussion

Another highly important job of some wildlife managers involves studying the impact of human activities on animal populations. These activities might include building a road through forest or locating an industry within wildlife habitat.

One technique used to evaluate this impact is called simulation. In simulation a wildlife manager uses a computer to model a population, such as deer in a woodland. The program uses rates of birth, death, and migration that the wildlife manager enters from previous records. Then the wildlife manager enters all events that might affect the numbers of deer through time, how likely these are to occur, and what effect each would have. When the program is run, it will give one possible fate for the deer population.

Method

In the following simulation, a checkerboard represents a wildlife refuge. The wildlife species (deer, for instance) are represented by two colors of chips (such as poker chips).

Rules of the Game

1. The members of your group will assume the role of a wildlife management team that has the following responsibilities:

    - Collect data (for several seasons) on the growth and decline of an imaginary animal population.

    - Plot the information on graph paper set up like Figure L10-1.

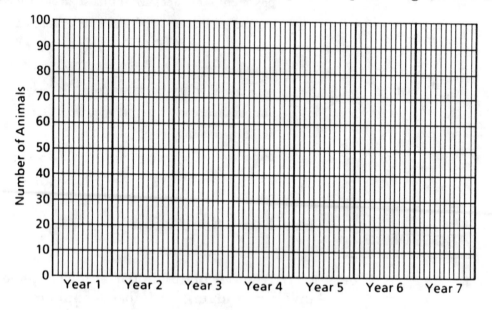

**Figure L10-1
Axes on
graph paper**

2. As you collect data through each season of mating, hunting, and migration (representing one round of play), you also will draw from a stack of cards (Impact cards). These Impact cards describe both chance and human-caused events.

3. After you have played two rounds of the game, you also will draw from a second stack of cards (Wildlife Management cards). The Wildlife Management cards describe ways that the wildlife population may be maintained.

4. After four rounds of play (four years), you will have collected enough data that you can begin planning your own management strategy. (You will base your strategy plays on what you see happening to the wildlife population on the board.)

5. The object of the competition among student groups is to see which lab group can best mainatin the wildlife population at its original size. Any group that eliminates its wildlife population is out of the game.

Setting Up the Board

1. Assign one color of chip to male animals and the other color to female animals.

2. Place all chips in the draw box.

3. Set up the board by drawing chips (without looking) from the draw box.

   - As you draw, place the chips randomly until all black squares are filled. (Black squares represent habitat with adequate food.)

   - Leave the red squares empty. (Red squares represent overgrazed habitat.)

4. Shuffle each of the two stacks of cards, and place each face down on the lab counter.

5. Before starting play, designate the black corner square of the board nearest the lab door to be your "key square." (The "key square" is referred to when you draw Impact or Wildlife Management cards.)

Rounds of Play

Round 1 (Year 1)

MATING SEASON

Any female on a black square can bear one offspring if her square is touching at least one square containing a male.

1. Draw chips randomly from a mixture of the two colors in the draw box, and place one chip on the same square and underneath each mating female to represent an offspring.

2. On a graph like the one in Figure L10-1 (which you draw in your ABC notebook), record the number of animals now on the board. (Don't forget to include the offspring in your count.)

IMPACT

3. Draw one Impact card from the top of the stack.

   - Follow the instructions on the card.

   - Unless it tells you otherwise, place the card face down in a discard pile.

   - If you run out of cards (any time during the game), shuffle the discard pile and draw from it.

4. Record (on your graph) the number of animals now on the board.

HUNTING SEASON

5. Hold the cardboard screen horizontally at arm's length and about one foot over the board so that the center of the screen is directly over the center of the board.

6. Slowly rotate the screen horizontally in either direction while your lab partner counts to three.

7. At the count of three, drop the screen onto the board.

8. Holding the screen firmly in place, remove from the board all chips (including offspring) that show through the notches in the screen. (Return these chips to the draw box.)

9. Record (on your graph) the number of animals now on the board.

IMPACT

10. Repeat Steps 3-4.

MIGRATION

11. Draw chips randomly from the draw box, and fill all empty black squares along the edge of the board.

12. Record (on your graph) the number of animals now on the board.

13. In rounds 2 and higher, remove all chips from red squares along the edge of the board, and return them to the draw box.

14. Record (on your graph) the number of animals now on the board.

IMPACT

15. Repeat Steps 3-4.

16. Rotate the board 90° in either direction. (The key square for the next round is the corner square nearest the lab door.)

17. Separate offspring and mothers by moving offspring to the nearest available square, red or black. (If no squares are

available, the offspring die of starvation caused by over-population.)

Round 2 (Year 2)

18. Repeat Steps 1-17 of the first round.

Rounds 3-4 (Years 3-4)

19. Repeat Steps 1-17 of the first round, but add the following at the beginning of each IMPACT section:

    • Just before you draw each Impact card, draw one Wildlife Management card from the top of the stack.

    • Leave all Wildlife Management cards face up (meaning they remain in effect) until you draw a "Funds Revoked" card (a card that gives opposite instructions).

Beginning with Round 5 (Year 5 and Beyond)

20. Leave the deck of Impact cards face down as usual.

21. Place each Wildlife Management card face up on the countertop so that all wildlife management options can be read by all members of the group.

22. Repeat Steps 1-17 of the first round, but add the following at the beginning of each IMPACT section:

    • Just before you draw each Impact card, choose (as a group) a management strategy from the Wildlife Management cards (one card only).

    • Apply management strategies as you continue to play.

Determining a Winning Group

23. Close to the end of the lab period (at a time designated by your teacher), stop the game and determine a winning group by counting the animals remaining on black squares. (The group that has the most animals on black squares at this time wins.)

Cleanup Instructions

Return all game materials to the place designated by your teacher.

Conclusions

1. List all events that caused the number of animals in the population to decline.

2. List all events that caused an increase in the number of animals.

3. Substituting your own data in the formula below, determine which year the birth rate of the species was the greatest?

$$\frac{number\ of\ animals\ born}{number\ of\ animals\ previously\ alive} \times 100\ \% = birth\ rate$$

4. Substituting your own data in the formula below, determine which year the death rate of the species was the greatest.

$$\frac{number\ of\ animals\ that\ died}{number\ of\ animals\ previously\ alive} = \times 100\ \% = death\ rate$$

5. In your opinion, what event was (or would have been) most responsible for the elimination of wildlife from the refuge?

Challenge Questions and Extensions

6. You are a wildlife manager. Based on everything you learned in the lab you just completed, plan the wildlife management strategy that would best maintain the species you studied in your lab game. (Your decision should consider the interests of miners, hunters, and animal conservationists alike.)

UNIT WRAP-UP ACTIVITY

Edison Power Company has developed its plans for the mine and power plant operation. The time has come for the county council to review the plans and decide whether to permit the plant to be built. The county council will hold a meeting to hear the positions of the citizens on this issue. Then, the council will vote on whether or not to grant a permit to Edison Power so the plant and mine can be built.

Hold the council meeting in your class. Members of the class will represent various positions and interests.

Each class member will play one of these roles:

- **"Pro" citizen.** These citizens focus on the ways that Richmond and surrounding communities will benefit from the mine and power plant operation.

- **"Con" citizen.** These citizens focus on possible harmful effects that the Edison Power operation could have on the natural resources discussed in this unit.

- **Edison Power representatives**. Several representatives from Edison Power will be present. They will respond to questions and clarify technical issues that may be raised in the discussion.

- **County councilperson.** Each councilperson will listen to the arguments of both citizen groups. Then they will cast their individual votes about granting the permit. They will also explain why they voted as they did.

- **Chair of county council**. This person will preside at the meeting. The chair will ask councilpersons for their votes and their reasons for voting that way. The chair will announce whether or not a permit will be granted Edison Power.

To prepare for the council meeting, each participant in the meeting should do the following:

1. Review all the Edison Power scenarios in the unit.

2. Evaluate each scenario in light of what has been learned about natural resources.

GLOSSARY

absorption – the process by which a substance takes up materials or
energy

acid rain – rainwater with a high acid content (pH <5.6) that results
from the reaction of water in the air with oxides of nitrogen and
sulfur. A major source of these oxides is burning fossil fuels.

acidic – the quality of water solutions that contain a high
concentration of hydrogen ions (hydrogen atoms with a positive
charge). Acidic solutions have a pH less than 7.

activated charcoal – a highly adsorbent form of carbon used to
remove contaminants from fluids

active solar – the use of sunlight to produce heat by exposing solar
collectors to the sun

adsorption – process by which one substance is attracted to and held
on the surface of another substance

adsorption capacity – the maximum amount of a matter that
can be attracted to and held on the surface of a given amount of a
substance

alkaline – the quality of water solutions that contain a high
concentration of hydroxide ions (molecules of hydrogen and
oxygen with a negative charge). Alkaline solutions have a pH
greater than 7.

aquatic – organisms whose normal habitat is water

aquifer – underground rock formation that contains water

ash – solids, including minerals, that remain after a substance is
burned

atom – the smallest form of an element that has all the properties of
that element

basic – alkaline; having a pH greater than 7

bedrock – layer of earth below topsoil and subsoil, usually solid rock

biodegradable – capable of being broken down into basic
components, usually by microorganisms

biogeochemical cycle – the processes by which a mineral is recycled through the environment

biological control – methods of controlling pests by introducing and encouraging natural predators or of reducing the harmful effects of a pest population in some biological way

boiler – a network of metal tubes in which water is heated to steam

calorie – the amount of heat energy needed to increase the temperature of one gram of water by one degree Celsius

carbohydrates – organic molecules containing carbon, hydrogen, and oxygen, which are used by organisms as a primary source of energy

carbon dioxide-oxygen cycle – the processes by which carbon dioxide and oxygen gases move through the environment

chemical bond – a force of attraction between two atoms that represents the storing of chemical energy

chemical formula – a description of a molecule that shows the type and number of atoms it contains

chemical reaction – changing of substances to other substances by breaking existing chemical bonds and forming new chemical bonds

chlorophyll – pigment used by plants to absorb sunlight during photosynthesis

clear-cutting – the practice of removing all vegetation in a given area, leaving the soil with no support

combustion – the rapid combining of fuel with the oxygen in the air; the process by which fuel is converted to heat

composting – the mixing of organic matter and other materials into soil to speed up their decay into nutrients

compound – two or more different elements joined together by chemical bonds

consumptive – the use of any resource in a way that harvests and destroys it

cultivation – growing or raising of organisms, usually vegetable crops, by domestic means

degradation – lowering the quality of a resource, that is, polluting it

depletion – using up of the total quantity of a resource

ecosystem – a biological community (plants, animals, and other organisms) and the nonliving parts of an environment with which it interacts

electron – a negatively charged particle found outside the nucleus of an atom

electrostatic precipitator – a device that traps particles in industrial waste gas by attracting them to electrically charged plates

element – one of the basic substances that make up the world and determine the characteristics of all materials

embryo transplant – transferring an embryo from the womb of its mother to the womb of another female in which it completes its development

emission gas recycler – a device that returns the exhaust gases of an internal combustion engine to the combustion chamber for more efficient burning of fuel

endangered species – a species with so few living members that it soon will become extinct unless measures are begun to slow its loss

energy – the ability to do work

erosion – loss of soil from an area because of flooding, wind, or rainfall

evaporation – the change of liquid water into water vapor

evapotranspiration – movement of water to the atmophere by evaporation from soil and surface water and by transpiration from plants

extinction – the dying out of a species

fertilizer – a substance added to soil to supplement its nutrient content

fluid – a substance that is able to flow. Air and water are both fluids.

food web – the network of food relationships in a community of organisms; the connections of who eats whom

fossil fuel – hydrocarbon compounds derived from the buried remains of organisms

generator – a machine that converts mechanical energy to electrical energy

genetic trait – a characteristic of an organism that is inherited from its parents

geothermal – superheated water and steam that are produced by volcanic action under the surface of the Earth

global warming – the current increase in the Earth's average temperature due to the effect of increased CO_2 levels in the atmosphere. These increased levels enhance the **greenhouse effect**.

greenhouse effect – the trapping of heat close to the Earth by CO_2 in the Earth's atmosphere

habitat – the physical environment in which an organism lives

heat – thermal energy that passes from a warmer object to a cooler one

heat content – the number of calories released for each gram of fuel that is consumed

herbicide – a chemical used to destroy unwanted plants

host – an organism that provides nourishment and/or shelter for a parasite or pest

humus – decaying organic matter making up topsoil

hybrid – an organism that inherits from its parents some characteristic that is different from either of the parents

hydrocarbon – a compound made only of atoms of hydrogen and carbon

inorganic – a substance that contains few or no carbon atoms in its molecules

integrated pest management – the planned use of multiple solutions to pest problems. Solutions usually include biological controls and some degree of pesticide use.

irrigation – the watering of crops with water pumped from underground or piped from surface waters

landfill – an area where wastes are buried between layers of earth

leaching – the loss of plant nutrients from the soil as they dissolve in water that the soil cannot hold

lignite – a brownish-black coal of low grade

limited resource – a resource that may be expected to run out some time in the future; for example, petroleum

loam – soil with an equal mixture of sand, clay and silt

mass – a measure of the amount of matter

mineral – an inorganic compound present in the Earth's crust. Rocks are usually combinations of minerals.

mixture – a combination of two or more substances that are not bonded chemically and that therefore retain their different chemical properties

molecule – the smallest possible form of any compound

nitrate – a form of nitrogen that can be absorbed from soil by plant roots

nonconsumptive – the use of a resource in a way that allows it to be renewed

nonrenewable – a resource that cannot produce more of itself and is not replaced by natural processes

nuclear fission – the process of splitting the atomic nuclei of elements such as uranium to release large amounts of energy

nuclear fusion – the process of joining together the atomic nuclei of elements such as hydrogen to release large amounts of energy

nucleus –the central core of an atom in which the atom's protons and most of its mass are found

ore – a natural mineral concentration that can be profitably extracted and refined

organic – a substance in which carbon atoms form the center or "backbone" of its molecules

overharvesting – destroying plants or animals of a given species faster than the remaining plants or animals can reproduce

oxidation – process in which a substance combines with oxygen. Some examples are combustion, rust, and the burning of calories in the body.

oxides – compounds of oxygen and other elements

ozone – an unstable form of oxygen that may break down into O_2 plus an extra oxygen atom. Ozone is part of the layer of the atmosphere that protects the Earth from the harmful ultraviolet radiation in sunlight.

parasite – an organism that lives on or inside another organism and damages it in some way, usually by causing disease

particulates – small particles of a substance, especially those suspended in a gas

passive solar – use of sunlight to produce heat by maximum exposure of a building to the sun

periodic chart – chart that contains all the known elements, arranged according to their chemical bonding characteristics

pest – an organism that has become so abundant that it is threatening the survival of some species that it uses for nourishment or shelter

pesticide – a substance used to destroy pests, especially insects

petrochemical – a chemical derived from fossil fuels

pH scale – a scale of numbers from 1 to 14 used to indicate how acidic or alkaline (basic) a substance is. A pH of 7 is considered neutral.

photosynthesis – process by which plants combine water and carbon dioxide in the presence of light and chlorophyll to make carbohydrates for food. Photosynthesis also releases oxygen to the atmosphere.

photovoltaic – generation of an electric current by exposing certain substances to light

pigment – a compound that absorbs energy in the visible light range

plant extract – substances obtained by separating and processing the juices of various plants

precipitate – a solid that forms when a substance will not further dissolve in a solvent

precipitation – any liquid or solid form of water particles that fall from the atmosphere to the ground. Rain, snow, sleet and hail are forms of precipitation

predator – an organism that kills and eats another organism called its prey

prey – an organism that is killed and eaten by another organism called its predator

primary nutrient – a nutrient required in large amounts by organisms; includes nitrogen, phosphorous, and potassium

productivity – the amount of bodily tissue that a plant or animal adds to its weight in a certain time period

products – substances formed during a chemical reaction

proton – a positively charged particle in the nucleus of an atom

quarantine – isolating materials or living things to determine if they carry parasites or pests

radioactive – a term used to describe a substance that emits streams of subatomic particles as it changes to a stable form

reactants – substances that combine in a chemical reaction

recharging – the process by which an aquifer is replenished; that is, by seepage of water through the ground

reclamation – the restoring of disturbed land to some usable state

recycling – returning a used product to a raw material that can be used to make a new product

renewable – a resource that can produce more of itself or that can be replaced by natural cycles; for example, trees

respiration – process by which organisms get energy from carbohydrates that react with oxygen. Respiration also releases carbon dioxide to the atmosphere.

scaling – the forming of a precipitate on some surface, such as calcium carbonate forming on the inside of a water pipe

scrubber – a device that removes gaseous pollutants from the exhaust gases prior to their leaving smokestacks. Electric utilities use scrubbers to remove sulfur oxides.

soil conditioning – modifying the soil to enhance plant growth

soil horizon – a layer of soil with characteristic features

solvent – a substance used to dissolve another substance

species – a group of organisms that share a set of characteristics, which are passed to their offspring

specific heat – the amount of energy required to raise the temperature of one gram of a substance by one degree Celsius

structural formula – a description of the arrangement of atoms in a molecule

subsoil – a layer of soil below topsoil and above bedrock. Subsoil lacks nutrients needed by plants.

surrogate – an organism that serves as a substitute for another

synthesis – the creation by humans of a new product from other products

temperature – a measure of how rapidly the molecules of a substance are moving about

threatened species – a species that is expected to become endangered if nothing is done to help renew it

tidal generation – obtaining energy from the change in levels of high and low tide

topsoil – the uppermost layer of the soil, usually greatly enriched in organic matter

transpiration – the process by which water vapor is released into the surrounding atmosphere by plant leaves

turbidity – a measure of the amount of material suspended in a liquid

turbine – an engine with curved blades that are rotated by a gas or liquid under pressure

ultraviolet radiation – a form of light energy, coming primarily from the sun, that is higher in energy than visible light energy

unlimited resource – a resource that is not likely to run out in the future; for example, sunlight

water cycle – the processes by which water moves through the environment, including rainfall, runoff, seepage, and various forms of evaporation

watt – a unit of energy equal to about 860 calories. Amounts of electricity produced are counted in kilowatts (thousands of watts).

weathering – breakdown of rocks into soil particles by natural physical or chemical processes